AF228899

BLUE ZEUS & FAMILIES

CLARE STAPLES
Founder of Skydog Ranch & Sanctuary

WILD HORSES OF SKYDOG

BLUE ZEUS & FAMILIES

Photography by JAMIE BALDANZA & STEVE RYMERS
Foreword by MATTHEW RHYS

TRAFALGAR SQUARE
North Pomfret, Vermont

First published in 2023 by

Trafalgar Square Books
North Pomfret, Vermont 05053

Disclaimer of Liability
The author and publisher shall have neither liability nor responsibility to any person or entity with respect
to any loss or damage caused or alleged to be caused directly or indirectly by the information contained in this
book. While the book is as accurate as the author can make it, there may be errors, omissions, and inaccuracies.

Trafalgar Square Books certifies that the content in this book was generated by a human expert on the subject,
and the content was edited, fact-checked, and proofread by human publishing specialists with a lifetime
of equestrian knowledge. TSB does not publish books generated by artificial intelligence (AI).

Library of Congress Cataloging-in-Publication Data
Names: Staples, Clare, 1964- author. | Baldanza, Jamie, photographer. |
  Rymers, Steve, photographer.
Title: Wild horses of Skydog : Blue Zeus & families / Clare Staples ; with
  photographs by Jamie Baldanza and Steve Rymers
Description: North Pomfret, Vermont : Trafalgar Square Books, 2023.
Identifiers: LCCN 2023014749 (print) | LCCN 2023014750 (ebook) | ISBN
  9781646012138 (hardback) | ISBN 9781646012145 (epub)
Subjects: LCSH: Skydog Ranch and Sanctuary (Or. and Calif.) | Blue Zeus
  (Horse) | Wild horses--West (U.S.)--Biography. | Wild burros--West
  (U.S.) | Animal welfare--West (U.S.) | Public lands--West (U.S.)
Classification: LCC QL84.22.C2 S73 2023 (print) | LCC QL84.22.C2 (ebook)
  | DDC 333.95/4160979493--dc23/eng/20230506
LC record available at https://lccn.loc.gov/2023014749
LC ebook record available at https://lccn.loc.gov/2023014750

Photographs by *Jamie Baldanza* and *Steve Rymers*. Additional photographs by *Shannon Phifer, Janelle Hight,
Steve Paige (American Wild Horse Campaign), Larry McFerrin, Laura Leigh* (Wild Horse Education), *Rayna
Zemel, Clare Staples, and John Wheland*.
Book design by *Katarzyna Misiukanis–Celińska (https://misiukanis-artstudio.com)*
Typefaces: *Miller Text, Minerva Modern, Nunito Sans*
Cover design by *RM Didier*

Printed in China
10 9 8 7 6 5 4 3 2 1

Dedication to Barkley

This book is for my boy Barkley, who lay at my feet
as I wrote it, made me slow down to find the time to finish it,
and has been by my side for the whole of Skydog.

Each photo reminds me of the invaluable part
he played in creating a safe haven for every wild horse
and donkey who stepped out of a trailer
onto hallowed ground, to be greeted by Barkley.

I will miss you every day, forever, my friend.
Go softly, sweet boy.

Additional photography by
SHANNON PHIFER, JANELLE HIGHT, STEVE PAIGE (AMERICAN WILD HORSE CAMPAIGN),
LARRY MCFERRIN, LAURA LEIGH (WILD HORSE EDUCATION), RAYNA ZEMEL,
CLARE STAPLES, AND JOHN WHELAND

foreword

by MATTHEW RHYS
New York, New York

FOREWORD

The image was jarring. Because both subjects
and setting were in such stark opposition.
Mustangs in a prison. Animals synonymous with
freedom, behind a high chain-link fence and razor wire.
This was the first time my great friend Clare Staples
and I met Sonny Jim and War Bonnet—Jimmy
and Bono as we came to know them. Two brothers.

They were exactly what you'd expect of Mustangs. Muscular in all the right ways and not too tall. Years of running barefoot over arid Nevada scrub had given them big, solid, sure feet. One word seemed most fitting for them overall: *hardy*. I loved them.

A few months prior, Clare had introduced me to her beautiful Mustang, Buddy, a golden palomino in every sense. I had enquired about Buddy's history and how a once free-roaming horse had come into her hands. It triggered a series of events that it had never occurred to me I'd see.

A handful of prisons and rehabilitation facilities in the United States offer and run Mustang training programs. In an effort to try to help alleviate the growing "incarcerated" Mustang population—held in corralled lots across the West—wild horses go to these locations where inmates prepare them to transition to domestic life. It is a therapy that works both ways. It is seemingly the perfect fusion of two creatures that have enormous trust and past trauma issues. Inmates usually have around three months to train the horses, and then they are sold at public auction.

Clare and I watched firsthand as one man, Thomas Smittles, had both Jimmy and Bono reacting to his subtle hand signals as he put them through their paces in front of the bidding public. Our decision was swift—I would buy Jimmy, and Clare and I would co-own Bono. Within weeks, they were Buddy's

↓ *With Clare Staples, Buddy, Jimmy, and Bono*

of Skydog Ranch and Sanctuary were born atop these Mustangs as we roamed the hills around the iconic Hollywood sign and Clare wrestled with how best to help the horses' plight.

Knowing Clare as I do, it is little wonder to me now how much of an enormous success Skydog has become, but the measure of its success is based purely on the salvation of wild Mustangs and burros. It is also little wonder that she has become such an icon herself in this incredibly difficult, and at times, brutal endeavor.

The personal reward, or more so (seemingly to me), the "worthwhile" element to this perennial challenge is seeing these incredible animals run free and be wild once again, where they belong.

A far cry from seeing them behind high, chain-link fence and razor wire.

neighbors at the Sunset Ranch in the heart of Los Angeles, a far cry from Carson City, Nevada.

In the evening, after a day's filming, Clare and I would ride all over the Hollywood Hills and the four thousand or so acres that made up Griffith Park, past such iconic landmarks as the Griffith Observatory—made famous by James Dean in *Rebel Without a Cause*. It was at this time, because of these three special Mustangs, that Clare started to take a vested interest in the wild horses of the Western United States and became (in the eyes of the BLM, anyway) a rebel *with* a cause. Early conversations

MATTHEW RHYS

preface

by CLARE STAPLES
Skydog Ranch and Sanctuary
Bend, Oregon

PREFACE

One of the questions I am asked more than any other
is just how a girl from England came to start
a sanctuary for America's wild horses. I am happy
to tell the story of how and when this all began.

Certainly my love affair with horses started as soon as I was born. My grandparents were Welsh and lived in a little village called Glynneath in South Wales. As a family, from as early as I can remember, we would make the car ride from Surrey, England, where we lived, to my grandma and grandpa's little house in the shadow of the coal mines. We would all sing as we crossed over the Welsh border, a song that began, "We'll keep a welcome in the hillsides, we'll keep a welcome in the vales," and my excitement started to build because I knew that scattered all across the Brecon Beacons mountain range were wild horses. I would fight my sisters for the window seat in the car and gaze out, hoping to see one.

I have no idea why seeing a pony anywhere—on the side of a wild hillside, in a field in Surrey, in the back of a trailer on the road to school—brought out in me an excitement similar to what I felt on Christmas morning. When I was old enough, I would walk alone through the nearby woods, which bordered my house, to a field where horses were kept, and I would lure them to the fence with an apple and then extend my long spider legs over their backs. No bridles, bits, or saddles. Nothing at all between me and those horses—just the thrill of being one with them, their manes tangled in my hands as I held tight.

As I grew up and spread my wings, I left home in the United Kingdom for America. I was mostly in search of the cowboys I had seen riding horses on television programs like *The Virginian, Bonanza*, and *Little House on the Prairie.* My favorite show had been *The Adventures of Champion,* which was about a group of wild Mustangs led by "Champion the Wonder Horse," and a little boy and his dog who would go see them. I would listen to the theme song in my mind to help me fall asleep, to dream about the wild horses and the lands they roamed.

Throughout my twenties and thirties, horses rode back into my dreams and real life as often as I could find them and an excuse to be near them. While I was in New York, I dated an Argentinian polo player and had the opportunity to travel with him to Palm Beach, Florida, and spend my days warming up his horses and hanging out in the barn with the grooms so I could be close to the horses I loved. I went on riding holidays with a group of friends who loved horses too. I rode around Central Park on a horse called Fury, rented at the iconic Claremont Riding Academy. I always somehow found a way to get back to horses.

It was when I moved to Los Angeles that the seeds of Skydog Ranch and Sanctuary were born. I came to purchase a golden Mustang named Buddy (I share his story, beginning on p. 10), and from the day he entered my life, I seemed to have a huge hand in the small of my back, pushing me away from the "glamorous" life I was living, toward something I can only describe as "my calling." Little did I know that I had found my life's purpose in my childhood passion. I look back now and realize that every memory and experience was preparation and training for what I would eventually undertake.

There is a paragraph at the start of Marguerite Henry's book *Mustang: Wild Spirit of the West,* that resonates deeply for me:

If God has a kind of plan for all of us, I like to think he coupled me with horses right from the start. It is not just my own Mustang that is part of me. All horses call to me. We sort of belong together. This could not just be an accident.

That is equally true for me.

Once I fell in love with Buddy, it seemed my innate interest in the wild horses of the American West only grew and grew. I felt like these horses had saved me, more times in my life than I could remember. It became apparent to me that every time I had gotten too close to the edge of danger and of losing myself, the horses had galloped in and given me my sanity back, grounding me in my first true love, and bringing me home.

But the evolution of my relationship with wild horses was not just a happy one. I went from that early love, to fascination, to outrage...and eventually, to horror, as I quietly educated myself in the facts of what was happening to Mustang populations, both in the wild and in captivity.

I quickly became determined to help the horses who had helped me right back. I saved some, learned more, saved some, learned more and more—until I felt strong enough and sure enough to dive right in, head first, with a solid intention to save *them all*.

I began with a goal to rescue seventy-five captive Mustangs, and then maybe retire to the wilds of Oregon, where I felt most at home and where I dreamed of them roaming free in a sanctuary, as close to their natural habitat as possible. I could sit on a windswept hill under the pines and watch babies grow up in their family bands. I'd be surrounded by the horses who had always given me joy.

With Buddy's herd

My reality now has surpassed that original dream, and what a blessing that is.

When I began Skydog Sanctuary, I came into it with no strong opinion on the issue of managing wild horses on public lands. According the United States Department of Interior, there are "400 national parks, 560 national wildlife refuges, and nearly 250 million acres of other public lands" that they manage. The agency within the US Department of Interior responsible for administering federal lands is the Bureau of Land Management (BLM). It has "oversight over 247.3 million acres and governs one eighth of the country's landmass" for the purposes of "multiple use." And it routinely "gathers" wild horses and burros from these public lands under the premise that otherwise, herd health is compromised and rangelands overgrazed due to overpopulation.

The BLM has a large network of holding facilities (what they call "off-range corrals") where the horses they have removed from the wild are kept, pending sale or adoption. As I began my mission to adopt Mustangs who had been pulled from the range in order to give them a life of freedom at the sanctuary, I started dealing with the BLM on a regular basis, as I spent a lot of time at the BLM corrals. Unfortunately, what I often saw in the wranglers who worked there was disrespect, cruelty, mistreatment, and disdain for the wild horses in their "care." It horrified me. This wasn't about the debate over whether or not these horses needed to be taken off public lands in the first place, losing their freedom and families forever, it was about what I perceived to be animal abuse once the horses were in the government's custody.

Time and time again I went to the corrals, hoping for cooperation as I strove to find ways to help the horses—in particular the seniors, those with special health needs, and the wildest ones who wouldn't likely adapt or transition well into domestic life. But it quickly became apparent that the staff at those corrals were not interested in any gesture of kindness. They were rude, disparaging, and condescending. They seemed to resent my compassion for the horses and the very idea of the sanctuary.

↓ *With Marvin*

With Scout

From the second Mustangs are born, they are targeted by people who can make money off them. From the helicopter pilots who are paid to chase wild herds from the air, to the ranchers charging to stockpile the horses on their land, to the off-range facility contracts, to the kill buyers, to those who "adopt" horses just to get BLM adoption incentive money, to holding facility staff willing to take backhanders from people looking for the best horses, and even some advocates and wild horse photographers who make money for themselves while calling what they do "advocacy"—through it all, in many cases, the horses' suffering is ignored.

And so I made it my goal to not only save as many wild horses as possible, but to help raise awareness and educate people about the very real problems within and systemic abuse allowed by the agency tasked with managing them. To change things for the better for wild horses, we need to change the way we approach what it is we value about their existence. I think Skydog Ranch and Sanctuary has made great strides in doing just that. Not only is our aim to educate the public and promote forever-home adoption of wild horses, we also recognize the social nature of these wonderful animals, and so have made it our mission to reunite families of Mustangs that have been rounded up and torn apart. One of my primary goals is to keep bonded horses and families together at the sanctuary. At Skydog, we believe that wildness matters, that family matters,

↓ *With Bobcat, Tank, and Joey*

and that there is a better way to manage Mustangs, when free on public lands and when taken into captivity.

I have a memory from when I was eight or nine. After school, my mother would drop me at a riding stable called South Weylands. I would usually just do chores like mucking, preparing feed, cleaning saddles, and grooming (I would only get about ten strokes of the brush in before I would stop and just breathe in the glorious smell of "horse," as deep as I could into my soul).

One crazy day I went along the whole line of stalls and unlocked all of them, pulling back the rusty bolts one after the other, and opening the doors so the horses could come out. Well, of course the stable manager called my mother to come and get me! My lasting image is of all those school horses, galloping across the green grassy fields, running and bucking with delight. My mother turned around as I was looking out the back window of the car with a true sense of joy in my heart and asked me why I'd done it. And I replied, "They really want to be free."

That memory only came back to me recently when my sister reminded me of the story. It further explained to me that, somehow, I was *meant* to do this—start a sanctuary to let as many horses as possible just be horses, and to "re-wild" as many as I possibly could. It is in honor of that little girl who already knew.

Horses are my greatest teachers. All my happiest times have been with them or on them or helping them, and I will find a way to keep saving as many Mustangs as I can, until I take my last breath, with a pocket full of cookies, hay in my gray hair, and the smell of a horse's sweet breath in my nostrils.

It has been my honor to help rescue every single horse you will meet in the pages ahead...and still I owe them more.

CLARE STAPLES

introduction

You might wonder why Mustangs are rounded up by the United States government and why they just can't "stay wild." Until the Wild Free Roaming Horses and Burros Act of 1971, brought in by President Richard Nixon to protect the less than twenty-seven thousand wild horses left at that time, there had been years of "Mustangers" and ranchers catching and killing wild horses to be processed for chicken and dog food.

IN the days of the Lewis and Clark Expedition in the early 1800s, there were between two and seven million wild horses roaming free in North America. It was a time in which ecosystems were in balance, and there was pristine biodiversity and wildlife. Sadly, both the state of the range and the numbers of Mustangs has been deteriorating ever since. Today, there are only maybe sixty-five thousand horses running wild.

It is estimated that wild horses occupy eleven percent of public land, while cattle graze just over fifty-six percent. Forested land is only around one percent for horses and thirty-eight percent for cattle. When BLM grazing allotments are determined, fifty to sixty-five percent is allocated for cattle. The wild horses, elk, deer, and other wildlife get what is left.

And yet the cruel answer to improving the health of range lands is always to remove the horses and leave the cows.

In the pages ahead you will meet more than forty Mustangs and their friends and families, all formerly wild horses, gathered by the government and held in captivity.

I and the team at Skydog Ranch and Sanctuary worked to find and rescue each of these horses with one intent in mind: to release them once again so they might live out their days in freedom, together.

A herd with many wild horse families
at Skydog Ranch & Sanctuary

*A tribe's first encounter with the horse
was often so momentous that the story
lived for generations. A Piegan elder named
Saukamaupee interviewed by a fur trader
in the 1780s remembered first seeing a horse
from the neighboring Snake tribe killed
during a battle. 'Numbers of us went
to see him, and we all admired him.
He put us in mind of a stag who had lost
his horns, and we did not know what name
to give him. But as he was a slave to man,
like the dog, which carried our things,
he was named the Big Dog.' Later, they
started calling horse 'elk dogs.' Other tribes
called them 'holy dogs,' or 'sky dogs.'*

—

David Philipps

*Wild Horse Country: The History, Myth,
and Future of the Mustang*

chapter

BUDDY AND BIRDIE

1

1

I want to begin with an ode to my very first Mustang, Buddy. Buddy's brand on his neck ages him as born in 2003, and we know he was rounded up at a year old, just a baby. On that day he lost his freedom, his family, and his wildness. He was taken to a BLM facility in Carson City, Nevada, ironically inside a men's prison, which then became his prison for a while. He was trained in an inmate program, the same one featured in the 2019 movie *The Mustang*.

Buddy didn't do very well being "gentled," and his first adopters returned him for being too wild, so he went through the three-month program again. For that he was named "Birdman," after the Birdman of Alcatraz, as he was in prison longer than the other horses.

At the time I was living in Los Angeles, living a fast life, a glamorous life, rubbing shoulders with movie stars and dating rock stars, trying to fill that hole in my soul. But something was missing, gnawing at my insides, and pulling at my heart.

One day I was taking my Great Dane Mr. Big for a walk in Beechwood Canyon, in the shadow of the huge Hollywood sign, and I passed a string of people riding horses. I followed them to a stable called Sunset Ranch. A few weeks later I was at a party with the television presenter Cat Deeley,

and she told me she had a horse at that ranch, and I should come riding sometime. And that is where it began. My love for Mustangs, and my first promise made and kept to a wild one. I saw Buddy one day at the ranch and told his owner that he was my dream horse. Shortly after she asked me if I wanted to buy him. So I did.

I came to see Buddy and rode every day, although he was pretty wild. As we wound up the trails, I noticed how he would often stop and gaze out at the view below him and sigh. Every time he did that, I got the crazy thought that he was saying, "Help my family. Find my family." That feeling I had got me started on this journey because I made my horse a promise— *that I would try.*

Buddy's wisdom and strength gave me purpose, happiness, filled the hole I had been trying to fill with people, places, and things for years. He helped me get sober as he gave me a reason to get up early and a passion and a purpose that he gently led me to. He saved me.

Buddy came with me to Oregon to start Skydog. He was the first horse to set foot on the sanctuary ground, and as a natural leader, he helped us settle the wild horses we took in our first year. If ever we needed to bring a horse in for some kind of care or evaluation, I would halter Buddy, and the others would follow him anywhere.

Buddy soon settled into "married bliss" with two golden mares who he chose as his own—Hunter and Little Girl. Both had been saved from kill buyers, both definitely strong and pretty enough for my boy Buddy.

Hunter was in a kill pen in Louisiana when I first saw her photo. Someone had tied a halter made from yellow bailing twine around her face to make her look "broke," but she was anything

but. Little Girl was a tiny champagne-colored palomino, so fine-boned and pretty that, to this day, she has remained the lead mare of whatever herd Buddy is in.

In the early days of Skydog Sanctuary, I went on a search for any other horses from the herd from Little Humboldt, Nevada, that had been Buddy's. I found three mares. But it was much later on, when I happened to need one more rescue for a "Giving Tuesday" promise, that I called up a kill buyer who had a golden Mustang mare. The man said that if I didn't take her, he was going to "chop her head off." I didn't know whether he meant it literally or figuratively, but I did know I urgently needed to rescue her and get her safe. I named her Birdie.

↑ *Buddy and Birdie*

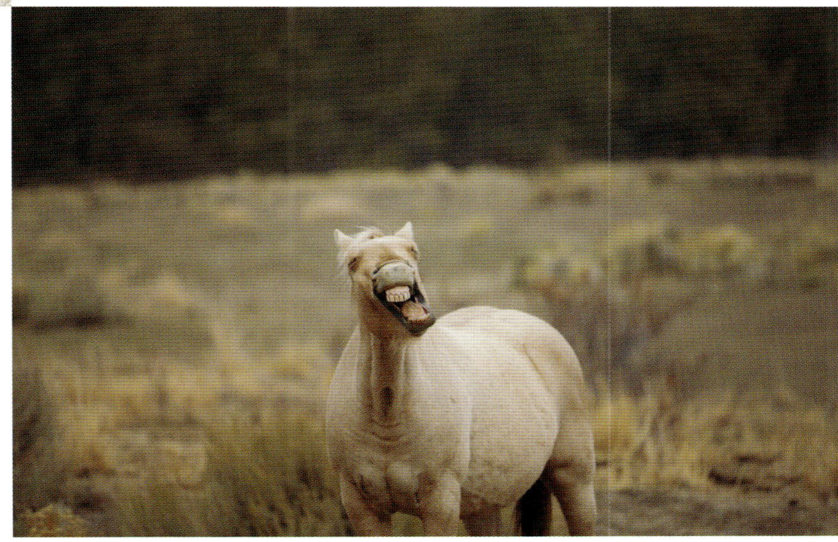

→ *Birdie yawning*

When I received a video of the mare from the quarantine manager, it struck me how much she reminded me of Buddy. But it was when she arrived at Skydog that I found out she was from the same herd and rounded up the same day as Buddy. Their brand numbers were only two digits apart!

Incredibly, I had named the mare Birdie without any reason that I knew. And Birdman was Buddy's original name. Coincidences are God's way of staying anonymous. Through Birdie, I got to finally give Buddy his sister back and help his family.

Promise kept.

We reunited Buddy and Birdie shortly after she arrived. I have seen a lot of family reunions now, but this one was so personal and immensely special. I often find the two of them together, even when they are part of a huge herd. He knows her and she knows him. They had been rounded up together and lost each other for seventeen years—a lifetime—but somehow, she had found her way back to her brother.

Families are so incredibly important to wild horses, and it is a travesty that they are torn apart the day they are rounded up. Not only have they have lost their freedom, they have lost parents, brothers, and sisters. At the sanctuary, when we rescue a Mustang, we then do all we can to find photographs of the horse's herd in the wild or the roundup so we can piece back together what was so wrongly taken.

 **skydogsanctuary**

So today's the day 🌼 We drove Birdie the short distance from the barn to the pasture Buddy was in. And within moments she was flying out of the trailer and into her new life. At first Bird didn't see another horse, she was stunned at the space and surroundings and taking it all in. But then Buddy called to her and she turned, saw him and ran across the pasture to meet him. And in that one moment of them meeting I felt so many different emotions it was goosebumps and heart sparkles and the biggest smile 😊 I had managed after all these years to bring a fragment of my Buddy's family home to him. It was one more promise I got to keep, honor and fulfill to a wild horse. Without Buddy there would have been no Skydog. He was the one who lit the flame in me to help these incredible horses. The years we spent together, the bonds we share, are just as deep as the ones between horses we now save. I know Buddy is proud and I know he knows we did good. So seeing Buddy gently greet his little "mini me," the tiny mare from his tiny homeland who was rounded up the very same day he was 17 years ago, processed two numbers apart, this was a day to celebrate and treasure. And another wild horse gets to be free and safe. Forever. #givingtuesday gave us all so much this year—thank you beyond words 🙌🐴🌼 Buddy got his family back after all these years.

Buddy and Hunter, grooming each other

Little Girl, running to Buddy

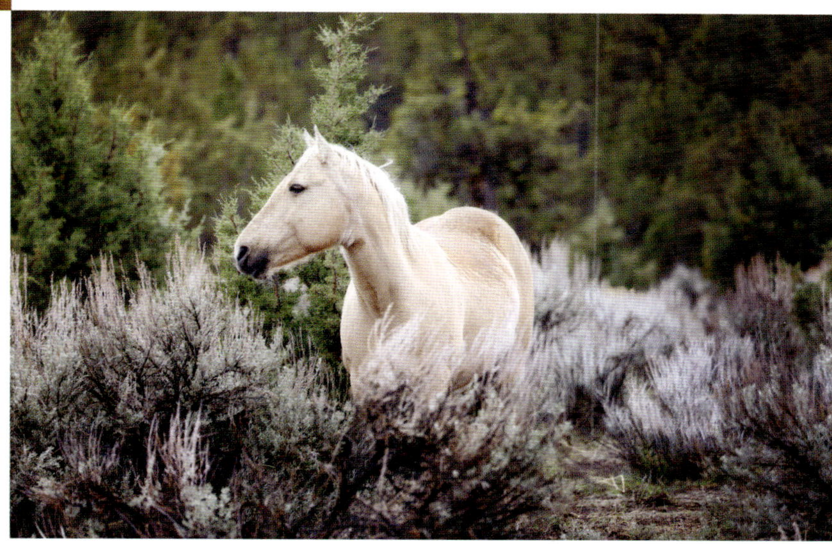

→ Buddy

We rescue Mustangs from a wide variety of different situations: from starvation, abuse, and neglect; from Mexican tripping rodeos, which traumatize horses mentally and often seriously hurt them physically; from kill pens and auctions; and from BLM holding facilities, either in an effort to reunite family bands or to take senior and special needs horses who are unadoptable. We will never stop fighting for them, and turning the light back on in their eyes by giving them back their freedom and wildness. What a gift it is to see these horses well and healthy, on land without a fence, as far as the eye can see. It will never get old, the thrill will never wear off, and my happiness is found in the peace in their eyes.

I am so grateful to Buddy for all of it. He is the reason I continue to reunite other wild horse families. I don't ride him anymore, but my love and gratitude for him is immense. I often go out to see him. He will always leave his friends to mosey over and say hello to "Mum." All the horses have respect for him—he has his pick of the hay piles! It's as if somehow they know he's the Lord of Skydog.

Wild horses! Just saying the words sets off
a stampede of images: echoing box canyons
and dusty blue mesas, hooves flying through
golden grass, the defiant scream of a rearing
stallion, heat waves rippling the distance,
speed and strength and cliffs and cactus
and dust and grit, lonely places
where big empty skies define the day
and coyote songs define the night, wild places
forever beyond the grip of civilization.

—

David Philipps

Wild Horse Country: The History, Myth,
and Future of the Mustang

chapter

BLUE ZEUS AND FAMILY

2

2

Somehow, this family reunion and rescue is the most meaningful to me, personally. It also is the biggest family we ever had the good fortune to bring back together, thanks to the bonds the mares and "children" had, which held them together throughout their roundup, and all the way to the moment we could release them once again in the care of their lead stallion, an incredibly striking Mustang named Blue Zeus.

Over the years I have seen hundreds of photographs, both amateur and professional, of wild horses on the range. I like looking at them, as they help tell the story of what it is we are fighting for at Skydog. Since the moment in Los Angeles when I heard Buddy tell me to "find his family, help his family," I have taken that to mean *all* wild horse families. And one of the most spectacular family bands I ever saw photographed was led by Blue Zeus on the BLM's Red Desert Complex in Wyoming. Many photographers traveled there to capture the multi-colored herd's wild beauty. When I saw their posts on social media, I would often leave a comment underneath, as this stallion had won my heart. I always said to people that as long as Blue Zeus and his family were wild and free, we had something to fight for and believe in. To know that such beauty still existed on our public lands was so important to me. Blue Zeus was my North Star, and he came to me in my dreams, pushing me on and forward with our mission at Skydog. His was a family like no other for me.

While we had rescued families before Blue Zeus's, they were not horses I had seen or known of when they were wild. I had only seen photographs of them once they were captured and corraled. Blue Zeus and his band were different. When I heard they were marked to be rounded up, I actually prayed for them to run far away and hide. I prayed for bad weather throughout the gather, hoping it would ground the helicopters and keep Blue Zeus safe.

But one evening I was checking the American Wild Horse Campaign "Gather Report" for the day, and there it was— an agonizing series of photos. There was no mistaking that Blue Zeus and his family had been captured. I stared at the images of a legend and icon of the West, losing his freedom, and his family, and my heart broke—shattered into a million tiny pieces. That night I vowed to do all I could to find Blue Zeus and as many of his mares and children as possible and bring them to Skydog.

This rescue was monumental in its effort and energy. Dozens of calls and emails to the staff at Cañon City, Colorado, where the herd had been taken. Befriending and bothering people by turns, trying whatever I could to get the attention of those who held Blue Zeus's fate in their hands. This was at the height of the COVID-19 pandemic, and access to any facility was mostly shut down; more so in this case, as the horses had been taken inside a high-security men's prison in lockdown.

I constantly reminded management at the facility that I was watching and waiting, and that the horses gathered from the Red Desert needed to be presented for adoption before being sent to "long-term holding"—a black hole for wild horses where there would be little chance of ever finding them again.

Finally I was told that there was to be an adoption event for just sixty-four of the senior male horses who had survived the gather and time at the holding facility. Janelle Hight, Skydog's Equine Director, and I made the trip to Cañon City to finally see for ourselves the faces that had been hidden away for almost a year, and to look into the eyes of the band stallions and see

Blue Zeus separated from his mares and babies

what twenty years of life on the range looked like in real life.

I was holding my breath the morning we arrived to see a mass of wild horses, packed into small corrals, unsettled, nervous, and worked up by the new environment. They huddled frightened in the back, trying to put as much space and distance between themselves and the humans at the event.

And then, there he was.

I actually have it on video—the moment I spotted Blue Zeus. I let out my breath, which I was still holding, and just cried. All the tension and fear I felt about him possibly not being there disappeared. I went to his corral and sat cross-legged in front of him in reverence, willing him to come home to

Skydog that day. I said a thank you to everything in the universe that had brought him safely to me on this day. I told him he was going to be alright and how good it was to meet him.

I was Bidder Number One as I was so early, and when it came time to write down my bid, I did so with a shaking hand: twenty-five dollars. Even now it seems hard to believe that was all his life and freedom had been valued at by man. I think I would have paid a million dollars to be able to take him home that day.

The bidding began; I don't remember anyone else being there in that moment. It was almost like an out-of-body experience. All I could hear was a man asking me if I still wanted the horse. I nodded, choking back a cry, and Blue Zeus was ours.

He traveled to Oregon in the caring hands of our hauler Carla Lays and in the company of the other old boys we could rescue that day (I share more about some of them on p. 206). In the morning, I was awake by dawn and

→ *Huckleberry, Iris, Phoebe, Flora, and Charlotte*

there to meet Carla and unload Blue Zeus. He was skinny and torn up, but he was alive and at Skydog, never having to move again, ever.

As the days went by, I spent hours just sitting near him, in the shade of an old juniper tree, just watching and talking to him. He was such a special soul; he was quiet and wise, considerate and thoughtful. He never felt the need to fight others for food and would just move slowly to another pile of hay if they came to get his. He seemed to feel no need to prove himself and content enough to just be back in a spot that resembled wildness, a far cry from the corrals packed with hundreds of other horses.

Over that time, I witnessed his palpable sadness, too. He would often be standing alone on a hill, gazing out at the land below him. I wondered if he was dreaming of another time, and I felt an aching longing for lost love and his previous life. Blue Zeus had a broken heart; I felt it much more from him than from any of the other older wild boys we had brought home with us from Colorado.

And so my resolve grew to go get his family. I had to find as many of them as I could and bring them home to him. It became my mission, stamped on my heart and mind, a promise I made to Blue Zeus like the one I had made to Buddy. I promised him I would bring his family home and that he wouldn't be alone anymore. I cried as I did so.

↑ Charlotte

Off I went, badgering the powers that be, insisting on an adoption event for all the Red Desert horses still stuck inside the Cañon City prison, unseen and unadopted. Finally I received an email saying there would be a small, private adoption event for those people who had asked to come into the facility, and the day was just around the corner.

I flew to Colorado Springs alone but determined. I couldn't sleep the night before the event and was up before dawn to drive to the prison. The parking lot held a few trucks and I recognized some faces. We were put on a shuttle to take us to the horse pens and then divided into groups to look at the mares and babies, the boys, the yearlings.

It had been raining and the pens were muddy, dirty, and slick. It was hard to wade through. I clutched the photographs I had of Blue Zeus's family to help me identify them. Pen after pen, I looked at every sorrel, every dapple gray, every black-and-white pinto. I felt his lead mare, who I called Nike for the pronounced large "swoosh" on her side, was the key because she was so distinguishable. If I could find her, I thought perhaps I would find them all.

On and on we went. At the second-to-last pen, as all the horses moved from one end to the other, a little colt skidded up to the

fence, stopped, and looked at me. I looked back at him and realized he was the mirror image of Blue Zeus's markings: the blue roan color, the high white socks, the white square on his neck. I wondered if it was some sign from above or just a reminder of who I was doing this all for. The colt's mother was a buckskin pinto, but her neck tag—which all the horses received when brought into BLM holding—was so covered in mud even the wranglers couldn't make it out that day. I made a note of the pen they were in, knowing that if I couldn't find the family members I knew from the photos, I would take him, in case he was Blue Zeus's son.

As I approached the final pen, I started to feel real fear that I wouldn't find any of them. Then, out of the corner of my eye, I saw a small group break away, and there was the black-and-white Nike with the distinctive mark on her side.

"The swoosh! The swoosh! There's the swoosh!" was all I managed to say, before realizing that right beside her was the sorrel, the gray, the babies, the yearlings. Blue Zeus's band was all still together, and their bonds had never been broken. They had somehow never been separated in the processing or holding procedures. A miracle indeed.

Blue Zeus and Nike

Blue Zeus and Nike 💙 🖤 Because we can never see too much of this boy. It's really hard to put into words how much I absolutely love this horse and his family. I hate the fact that he was rounded up and lost his freedom but every single day I get to share time and space with this boy and his awe-inspiring, take-your-breath-away beauty. The love he shows his beloved mare Nike, as well as each one of his children and mares is so spectacular. Would that we all knew a love so pure and deep. Separated for a year, it was a monumental task to find and adopt them all and then bring them home to reunite in Oregon. And their rescue has raised so much awareness for the simple fact that these horses have family they love and cherish and deep bonds that need to be respected, acknowledged, and revered. But this morning, high on the side of a hill, in view of Sheep Rock, all is well in our world 💙 🖤 Thank you so much!

Blue Zeus

I headed into the adoption office, clutching a small piece of paper with each tag number written carefully on it, and I sat down to fill out my paperwork. It was a crazy moment as I carefully signed each adoption form, my hand shaking and a lump in my throat, realizing what the sanctuary had managed to achieve for Blue Zeus. In the end the credit lay with Nike for showing herself so perfectly to me and for keeping her family together and in one piece for so many months. Nothing proves the familial bonds these horses can have better than that day. Penned with dozens of other horses for over a year, they had stuck by each other and never left each other's sides. And now they never would.

We were allowed to take the first four horses home sooner than the mothers with babies—I named all the mares after Roman and Greek Goddesses to match Blue Zeus's triumphant name: Nike, Gaia, Nyx, Phoebe, Juno, and Flora. Then, the actress Kristin Davis had reached out to ask us to try to save one more mare, and the way all the hauls worked out, we had one compartment free with room for another mare and baby to travel to Oregon. So I asked for the buckskin pinto mare and the cheeky colt who looked so much like Blue Zeus, and I named the mare Charlotte after Kristin's character on *Sex and the City*, and the colt Huckleberry.

It was a crisp, brisk winter morning when we brought Blue Zeus down to the Elk Barn, which adjoined the fifty-acre area where his family were turned out. As the sun came up over Sheep Rock, the gates were opened and out came Blue Zeus at a run, head so high and proud, chest expanded, mane and tail flying as he headed up the hill to his family.

And then what a dance they went on, following Blue Zeus in a procession, before he turned and greeted each one of his mares before leading them off over the hill to a more private spot, away from our cameras. It was glorious, the highlight of my endeavor to educate people about wild horse families with the most beautiful example. But more than anything, to be able to bring the thing that Blue Zeus missed and loved more than anything, home to him, became the greatest joy of my life.

Nobody can ever dull that incredible achievement. Nobody can ever diminish our happiness at getting to see this family grow up together, live as one, their dynamics a daily masterclass of the bonds horses share.

The Dodo, a digital publisher focused on telling animals' stories, did an incredible seven-minute movie about the journey I had been on to find and reunite Blue Zeus and his family, which they premiered on Christmas Day. And from that came a feature film documentary that I hope will help bring real lasting change for wild horses and increase and amplify the voices already calling for an end to their mismanagement. I think Blue Zeus will continue to inspire others to reunite horse families and allow people to see that horses are not livestock but deeply feeling and meaningful beings who deserve all the kindness and respect humans can offer them.

*With all due respect to our official icon,
the eagle, he of the broad wingspan
and the ability to see across great distances,
of patience born of the ages and of majestic
flight, it is really the wild horses,
the four-legged with the flying mane
and tail, the beautiful, bighearted steed
who loves freedom so much that
when captured he dies of a broken heart,
the ever-defiant Mustang that is our true
representative, coursing through our blood
as he carries the eternal message of America.*

—

Deanne Stillman

*Mustang: The Saga of the Wild Horse
in the American West*

chapter

SAGE AND JUNIPER...AND CASSIDY

3

3

I was contacted by one of the adoption staff at the BLM corrals in Burns, Oregon, about two mares who had been seized after a compliance check at their adopter's home. The mares had been adopted out together, but then neglected and treated very poorly. They had been brought back to the corrals and were again up for adoption, this time online.

The staff told me one of the mares had bids but nobody wanted her sister who wasn't as pretty and was partially bald from rain rot due to the neglect they had suffered. I was told how very bonded the mares were and asked if maybe I could bid on them both to bring them to the sanctuary so they could stay together.

I did just that and won the sisters, named Juniper and Sage, and we brought them to Skydog. Both had been traumatized by the training process at their adoptive home and were horribly scared of any contact with people. To this day, Juniper remains, without a doubt, the wildest and hardest horse to handle at the sanctuary. We don't even try to push her into the chute for wellness checks as she will do something crazy to avoid being handled or touched. She is happy being wild, and we respect that choice.

We turned Juniper and Sage out with Buddy's small, but growing, herd. And then, the funniest thing happened: our little mini mule Cassidy, who had been saved from a kill pen, went under a fence, leaving the donkey herd he had been turned out with to run with the wild horses and "claim" Juniper and Sage as his own.

Early on in Skydog's evolution, I had no idea what a "kill pen," a "kill buyer," or the "slaughter pipeline" for horses was. Cassidy, and two donkeys called Frankie and Johnnie, and a Mustang named Love, were the first rescues I made from a kill pen. I had to teach myself how to bail the horses, how to find a hauler to take them to the vet, how to find a quarantine person I could trust with the care of my rescued equines, and then how to get them all home. It was a fast and furious education—lives were depending on me, and I couldn't afford to get it wrong.

Buddy and Cassidy

skydogsanctuary

A sight for sore eyes 💙 A waterfall of rescued horses and donkeys and a mini hinny. Racing down the hill following their well-worn path on land they know and love. They are bonded now in all they went through before they made it here and no longer run from a helicopter but toward people they know and care for. This is what we all dream about for our wild horses and burros. A life free from fear, abuse, servitude, worry, empty bellies, separation, and cruelty. Here they can be horses again and just be. Live. Love. Run. Relax. And it's beautiful. We will never stop working for more of this for more of them. We will be out in all weather and up with the sun and to bed as it sets. Tired after a day's work caring for these incredible souls. But getting up the next day to do it all again. This my friends is a soul-soaring, heart-stopping, awe-inspiring sight. And you help us do this every single day. Thank you 🙏

→ *Sage*

Many people in the United States and around the world have no idea that every year thousands of horses, donkeys, and mules are sent across our borders to slaughter in Mexico and Canada. For these animals, the journey there and the way they are killed when they reach their destination is brutal and cruel, and we have worked for years, lobbying the US government to pass the Save America's Forgotten Equines (SAFE) Act, which is a piece of legislation that would stop the transport of equines over our borders to be killed for their meat and hides.

Every two years we make a movie to help raise awareness about this legislation, encouraging people to call on their state representatives to co-sponsor and support the bill, but so far, we have not made much headway in even getting it to the floor of the House of Representatives for a vote. Our kill pen saves, like Cassidy, stand as ambassadors for ending the brutal trade in horseflesh. As a member of the first group we ever saved from slaughter, it was incredible to see another chapter of his story play out—this part, a love affair.

After Cassidy decided that he was a Mustang in spirit and sensibility, we never saw him without also seeing Sage and Juniper close by. He stayed as close to them as he could, alternating duties between being their protector and mighty stallion, and being Buddy's lieutenant in the band. It was quite the sight to see this tiny boy, behaving as if he were the leader

and an equal with all the bigger horses in the herd. He had his mares, and he was second in command—brave as a lion and wild as the wind.

We lost Cassidy in the winter of 2022 to colic. It's hard to imagine a Skydog without him, as he touched so many hearts and people.

Sage and Juniper still live with the big herd and are still close to Buddy. Sage is the more adventurous of the pair and sometimes leaves Juniper to go on adventures with another herd, led by a horse named Phoenix (I share his story on p. 134). Sage has quite a few mare friends in that group, too, but she always comes back to her sister, putting family before all other bonds.

*When it comes to wild horses, their beauty
strikes us most, but horses possess more
than prettiness. The appearance of Mustangs
on the landscape in the West can be beautiful,
of course. But it is more. The site and
the experience cross over into the sublime;
the spectacle of wild horses, surrounded
by scenes of rugged topography
can produce a sense of reverence so strong,
it feels almost unnerving.*

—

Chad Hanson

*In a Land of Awe:
Finding Reverence in the Search
for Wild Horses*

chapter

BEAR, GOLDIE, AND AERIAL

4

4

AS with Blue Zeus, this story started with a photograph. In this case, a blurry, fuzzy, faded photograph of three horses standing in a kill pen. Three horses: a golden palomino mare with a white mane, a jet-black mare with a white blaze, and a liver chestnut boy with a blaze and white socks. They were standing close together, as if clinging to each other for comfort in the strange world they had been dumped in. And the most unusual part was I was informed the chestnut was still a stallion, which is practically unheard of, as the BLM gelds all male horses before adopting them out.

And so, I decided to help them, little knowing that this group, some of the first Mustangs rescued by Skydog, would help us learn so much about horse families and their bonds.

We bailed the three, who I called Bear (stallion), Goldie (palomino) and Aerial (black), and they went to a quarantine in Texas to wait until we could bring them to the sanctuary. While in quarantine, we had Bear, who was eight years old at the time, gelded to make it possible for him to be released at Skydog.

The three Mustangs reached us right before winter, and we watched as the two mares got ever bigger, rounder, and wider, leaving us in no doubt that they were pregnant with what would be Bear's last babies.

The veterinarian was at the ranch one day, and I asked her to look at Aerial because she was so huge I was sure she must be about to have her baby. The vet explained that it would be unusual for a Mustang to have a baby in January as usually foals are born in the spring and not to worry. The very next day Aerial had her baby (Whisper) in six feet of snow and minus six degrees overnight. He was the fluffiest colt I had ever seen, and I learned baby horses will grow excess coat, depending on the outside temperature.

Goldie's baby, Mariah, was born about six weeks later—a beautiful sorrel with a crooked white blaze to be a friend to fluffy gray Whisper, her half-brother. The mares were amazing mothers, and even though they were wild as the wind, they were happy to let us enjoy their babies. I had never had a foal before and couldn't get enough of burying my cold hands in their thick fluffy coats, sharing space and breath with them. It was a magical time.

Bear was the most incredible father to the babies. He would "run drills" where he would wake his children from their naps and have them run with him and his mares around trees, dodging buckets, from one side to the other of their pen at a full run. I understood this is what they would do in the wild if they sensed a predator nearby—the stallion would gather up his family and take them to safety. Bear was teaching the babies how to survive in the wild, and it was spectacular to watch.

Bear spent a lot of one-on-one time with his son, Whisper, too—sparring, wrestling, making him move his feet, teaching

Bear

 **skydogsanctuary**

Nirvana – (noun) any place of complete bliss, delight, and peace ☮ synonyms: Eden, Shangri-la, Heaven, Paradise, promised land. Nirvana is a place of perfect peace and happiness, like heaven. Just some of the definitions of the word and I often think it must look a lot like this. Sanctuary means a place of refuge or safety and surely that's what this is for these horses and burros….and for me and the people who work and live here too. With the green grass and rolling hills, streams, and Sheep Rock towering above us all – this is my heaven and place of peace and I hope it is for you too ☮

his boy how to one day win mares and fight other stallions for them. It was a privilege to watch the strong bonds Bear had with his mares and developed with his two children. It was my first up-close lesson in understanding how deep a love these horses feel for family; it taught me well and informed many of the Skydog rescues to come.

Bear had a thick, clumped baseball bat of tail hair, which he couldn't lift or swish to ward off flies. He was wild, so we arranged for a trainer to come to the ranch and work with him so we could relieve him of the matted mess and he could grow healthy, fresh tail hair. Goldie also gentled down a little—enough for us to halter and handle her as necessary—but Aerial remained as wild and spirited as the day she arrived at the sanctuary. Looking at her terrified expression in the photos of her being processed at the kill pen where we found her, we understand why.

Bear, Goldie, Aerial, and their babies went on to join other kill pen and adopted mares out on the sanctuary (I will tell you about a few of them in the pages ahead), and Bear took all the mares and foals on as his own. Early on, we had no idea how many mares in the slaughter pipeline are bred so they weigh more and fetch a higher price. It is astonishing the cruelty that exists in the underbelly of the horse industry.

To this day, Bear remains the patriarch of this band, and particularly close to his two mares and children. Often you will see Goldie and Mariah sharing breakfast, and Aerial and Whisper the same. It always makes me happy to see them together.

"

*Almost every tribe immediately yearned
for horses, dreamed of them, sang of them,
painted them on Canyon walls, named moons
in their annual calendar after them, and
welcomed them into the cultures so completely
that before long they were sure the horse
had always been there. The Apache said
the Creator made the horse, using lightning
for its breath, rainbows for its hooves,
the evening star for its eyes, crescent moons
for its ears, and a whirlwind for its power
and speed. The Navajo said that every day
the sun god rode across the sky on a turquoise
Mustang with a joyous neigh.*

—

David Philipps

*Wild Horse Country: The History, Myth,
and Future of the Mustang*

"

chapter

PRISCILLA AND DALTON

5

5

⤢ *Priscilla*
⤢ *Baby Dalton*
→ *Dalton*

Shortly after rescuing Bear, Aerial, and Goldie, we were alerted to two more Mustangs in a kill pen in Colorado. They were called Swayze and Priscilla. She was a pretty tri-colored pinto who we discovered, once she reached the sanctuary, was not a Mustang at all (we had been misinformed when we were told she had a brand). Nonetheless, we weren't going to send her back.

As time went on, we noticed Priscilla's belly getting rounder, and so we moved the mare to the barn and made a stall ready for her and what we could now see would be a baby. She was the size of a barn before she finally gave birth to a beautiful colt we named Dalton for Patrick Swayze's character in the movie *Roadhouse*. Her gelding companion Swayze had taken such good care of her on the road from the kill pen to Skydog, it seemed fitting.

Sadly, Swayze did not take kindly to Dalton; it seemed fatherhood was not to be, so we added Priscilla and Dalton to Bear's family band, and we waited and watched to make sure he would accept another baby. Miraculously, he did, and Priscilla now had "aunties" in the other mares to help her, and a strong father figure to help take care of her and her son. Bear went straight to drilling Dalton as he had his own foals, and teaching him to spar and wrestle and run.

And Bear's family was about to grow once more when we said "yes" to two more mares and foals, all the way from a failed sanctuary in South Dakota.

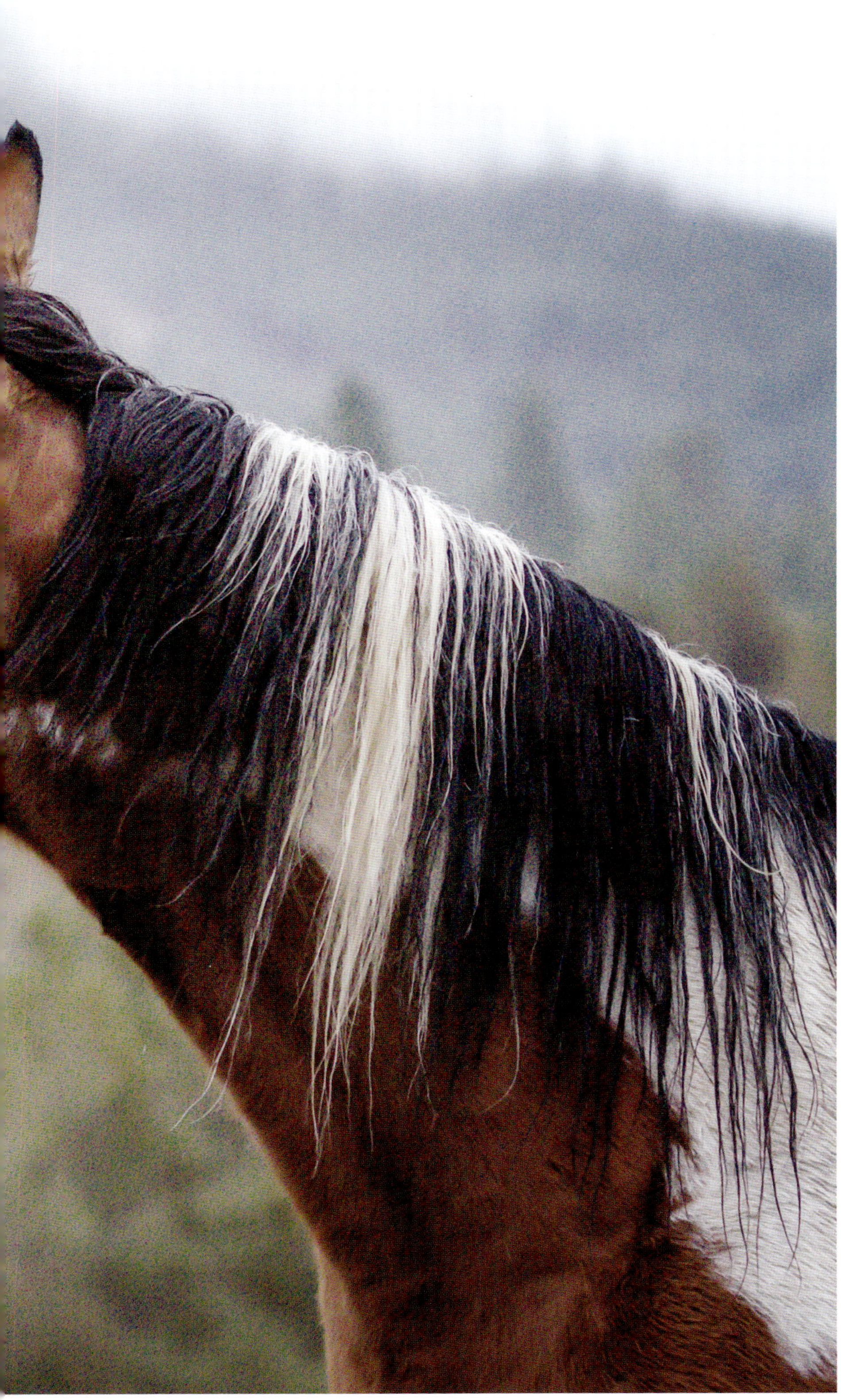

skydogsanctuary

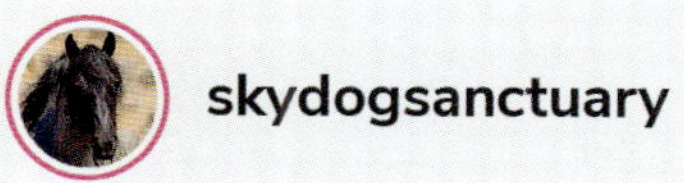

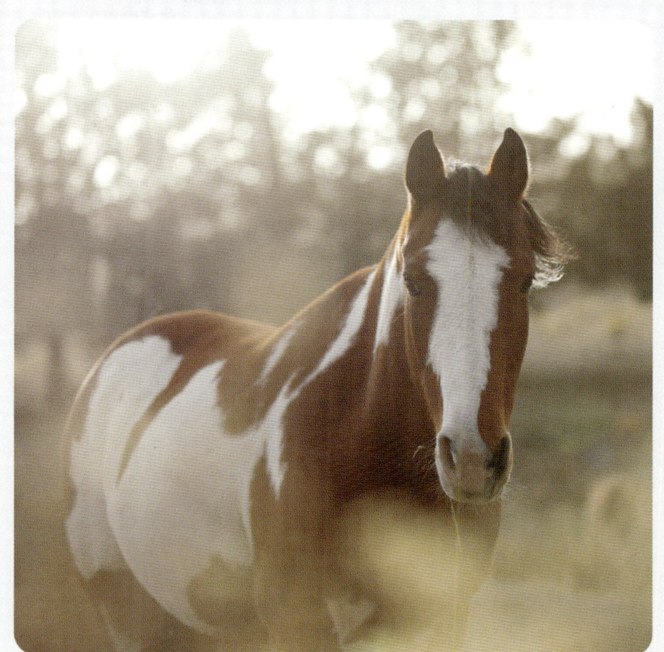

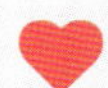

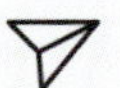

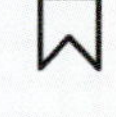

And now they are here, with the sun shining through the trees as it rises, their bellies full, friends surrounding them. And suddenly all is well. The nonsense of the outside world melts away and this is everything. Rescued horses, a Great Dane, and good people. Surround yourself with this and the noise and mean people can never win. Here is the peace, the proof, here is the joy and the reward that makes it all worth it. We rescue them and a hundred times they have healed and rescued me back. Plucked me from the overthinking, the self-doubt, the inner critic...and the outside ones too.

This is the great paradox of the horse.
It possesses a wild spirit but serves as
the greatest helpmate this country—and
all of civilization—has known. Other wild
animals have been pressed into service or
entertainment, but it is only the horse...that
consistently moves back and forth between
there and here, horizon and corral,
range and rodeo, inspiring centuries of song,
art, literature, and worship and stirring
passions that have wreaked havoc
in everyone from King Solomon
to the ancient Greeks to cowboy poets.

—

Deanne Stillman

Mustang: The Saga of the Wild Horse
in the American West

chapter

WINTER, BLIZZARD, SOOTY, AND SHELDON

6

6

When the Skydog team first saw the whistleblower images and first-hand reports of what was happening at a sanctuary in South Dakota, it was shocking and incredibly sad. The sanctuary had allowed horses to breed, and the population on the premises had exploded from around three hundred horses to over nine hundred. That winter the temperatures in that region were horrifying, too, with a negative-fifty-degree wind chill as the rescue effort began. An extraordinary group of determined women went to the site once the county officially seized the horses and began the monumental task of sorting them, arranging for the ones too far gone to be euthanized, and finding other sanctuaries to take the rest. It was the biggest rescue I had yet been a part of.

One of the ladies on the ground was Barbara Rasmussen, a dear friend I had worked with to rescue a number of Mustangs from kill pens. I donated personally to help with the transportation of some of the horses in South Dakota to new facilities, but with the sheer numbers and so many others taking horses, I waited to see what else might be most needed.

With unwanted horses, it often feels like the well-known story of the boy and the starfish. You can find it told a few different ways, but it goes something like this:

When hundreds of starfish are washed up on the sand who will surely die, a young boy starts picking them up and placing them back in the ocean, one at a time. A man comes by and tells the boy there are too many starfish, and he will never be able to save them all. The boy picks one up and puts it gently back in the ocean and says, "I saved that one."

That's how I try to look at the horses we rescue. I can't fathom the number of horses in need, the Mustangs in BLM holding facilities, the huge numbers crammed in kill pens, the neglected and unwanted across the country. I can only hold space for those we are able to save and know that to that one horse, it is his entire world and life. Otherwise, it would all be too much to bear.

So when Barbara reached out and asked if Skydog could take two mares who were dear to her, both with foals, I said yes. On a very cold day, the four Mustangs were loaded on a trailer and began the journey all the way from South Dakota to Oregon. I worried for them but told myself all they had to do was make it home. I named the first mare-foal pair Winter and Blizzard in recognition of the extreme weather, and the other two Sooty and Sheldon.

The four arrived safely, and we unloaded them as it snowed gently. The mares were *wild*. Wild *wild*. But the babies were sweet and curious, and we made friends easily. (We later learned that Blizzard the filly had fallen in a water trough in South Dakota, and after some kind people helped her out, she'd decided humans were her friends.)

For the first couple of months, we kept the four inside, in the warm barn. We dewormed them carefully, as they all had heavy worm loads, and dealt with other health issues. Little by little they

skydogsanctuary

To have their dignity, strength, and spirit back. I spent a life searching for happiness in places where it was not sustainable—but this gift we are being given to help these worthy animals out of a dark place and back into the light and love does more for me and all of us than anything material. Being of service, living a life helping others, it really is a purpose-driven life, and we are so grateful you are a part of it. You all are these horses' guardian angels and saviors, and we will never let them drop through the cracks again.

Mariah and Swift

WINTER, BLIZZARD, SOOTY, AND SHELDON

→ *Winter*

emerged, transforming from the dirty, sick animals who arrived that cold day to the beautiful horses we knew they were behind that rough exterior.

Winter came to us pregnant, and when her new foal arrived, we decided to add them to Bear's family to see if he would take on another couple of mares and three new babies. Well, he did, and was a superb father to all. It was beautiful to see the entire group of mamas and babies have a leader and a strong protector who doted on his family as we grew the sanctuary. Bear truly takes the responsibility of being a band leader seriously and to heart.

We brought the youngsters Sheldon and Blizzard to the Malibu facility so we could introduce them to more people and so they could be ambassadors for the Mustang cause. Very tragically, Blizzard was kicked by another horse while there, shattering her leg, and she had to be humanely euthanized. It broke all our hearts as she really was the first horse our Skydog team had loved and then lost. Thankfully, she didn't suffer; our veterinarian was with her in record time. And we were consoled by the fact that her newborn sister, Wilder, was still running free with her mother and family.

When you love these horses as we do, the losses can be so hard to bear. Each one is a different strike on the heart, and some seem so unfair. But "on we go"—that's the expression we use to carry us forward. At Skydog, we have so many other horses depending on us we have no choice but to keep going. And the horses we have

↑ *Sooty*

lost are remembered and honored in all the ways we can.

Nowadays, the whole of Bear's band is part of the big Skydog herd and running free on the main portion of the upper ranch in Oregon. It is quite the sight to see over sixty horses running in for breakfast in the morning and to remember the awful conditions they were all in when we got them. Saving family units seemed to happen accidentally at the start of our Skydog mission, and later, very deliberately, as we went about piecing groups back together after roundups. Each one is different, but all of them prove the bonds and the love horses feel for each other. We will always keep doing all we can to keep them, and others, together.

And on we go...

*Wild horses.... They are freedom.
They are independence. They are the ragtag
misfits defying incredible odds.
They are the lowborn outsiders whose
nobility springs from the adversity of living
a simple life. In short, they are American.
Or at least they are what we tell ourselves
we are, and what we aspire to be.*

—

David Philipps

*Wild Horse Country: The History, Myth,
and Future of the Mustang*

GOLIATH, RED LADY, EMBER, AND BODHI

7

7

When I started Skydog Ranch and Sanctuary and took a giant leap of faith into the unknown world of wild horse rescue, I had no idea of the importance of family to equines. Even though the first wild horse I had known (Buddy) had told me in no uncertain terms that I was to help his family, I didn't know enough about horse herd dynamics to truly and meaningfully comprehend the familial bonds Mustangs form. My understanding grew with each group of horses we brought to Skydog.

A friend of mine contacted me one day, distraught and heartbroken. She told me a Mustang known as Goliath had been captured with his family, and would I consider rescuing them. He was twenty-six years old and had lived wild and free on the range his entire life. Now, he was confined to a BLM holding pen in Utah.

I agreed to do what I could, truly not understanding the implications of that promise. I simply wanted to help someone who obviously cared very deeply for this horse.

I started off by contacting the BLM and asking if they would consider *not* putting Goliath as available on their online adoption site due to his age and size and wildness. All of these factors meant he was way better off being returned to the wild, and if that wasn't possible, given to a sanctuary. My pleas were

refused as the BLM did not want to set a precedent. I took a deep breath and announced we would bid on him so we could take him to Skydog. Never again will I make that mistake. I had no idea that there were people out there that would do all they could to make sure he didn't get to live the rest of his life at Skydog. I was utterly blindsided by the countereffort, and sadly could do nothing but bid as high as I could go and pray I won.

At $6,800, we were at last the winners. It was the most we ever paid for any horse, before or since. It was a hard lesson, and we learned never to let anyone know which horses we are interested in and to keep all plans secret until a horse is safe at the sanctuary.

We traveled to Delta Wild Horse and Burro Facility in Utah to collect Goliath and bring him home to Skydog to live a life as close to wild and free as possible. What he didn't know, what nobody knew except us, was that his mare, Red Lady, who was quite possibly pregnant with their last baby conceived before the roundup, was already safely in Oregon, waiting to reunite with her stallion.

We released Goliath on a beautiful, snowy day, and he took off to the other side of the fifty-acre pen he was in to get as far away from humans as he possibly could. But then we opened the gate and out trotted Red Lady, looking spectacular in the winter sun.

Suddenly Goliath sniffed the air and began the most glorious run imaginable across the snowy hills to his Red Lady. He sniffed her belly, and we instantly knew that she must indeed be pregnant, a fact we suspected but weren't sure of. He greeted her and then escorted her away, high up in the trees where he rolled to rid himself of all the smells of the holding pen he had been in

skydogsanctuary

For anyone who believes horses need a job I wish you could truly see the beauty and perfection of this. Wild horses grazing and slowly moving across the wide-open space, as they head for the water hole as the sun sets behind them. Being a herd and a family and all that complicated dynamic encompasses is deeper than a job. Their close bonds and obvious love for each other is palpable and moving. Bodhi and Magnolia were running and playing as youngsters do as Grace watched over them. Madison alert and attentive while Red Lady and Goliath munched down to just above the roots the grasses which grow wild here. Even as they poop they pass seeds and new life for this ground to renew itself. They are a part of the history of this country, part of the landscape and the horizon, part of our soul and spirit moves with them if you have a heart for them. Let's not lose this for future generations. This is important, this matters. Wildness matters and so do they. Goliath and his family are the heart and soul of this sanctuary, and in a few days this gentle giant, this living icon of the West is going to get one more big surprise. For the rest of this week we will be reminding everyone of this family's story, history, and how they came to be at Skydog and then .... we shall see 💙

GOLIATH, RED LADY, EMBER, AND BODHI

for three months. Red Lady never left his side again. And almost three months later, she gave birth to their baby, Bodhi, who was a gorgeous pale sorrel with a white tail that he still has traces of today. It was beyond the perfect ending to a love story.

On the day Goliath arrived, I sat on that snow-covered hill and told him that his roundup would not be in vain and that his capture would raise awareness for all the other Mustangs who weren't as lucky as he was to be with part of his family again. I decided to make a short film about him, using footage from kind photographers and others who had taken video of his herd in the

⌄ Red Lady and Bodhi
⌄ Goliath

→ *Red Lady, Ember, Goliath, and Bodhi*
↓ *Red Lady and Ember*

wild, who were all so incredibly generous in allowing us to use their material to help Goliath's family, and other wild horses, in any way they could. To this day, the story of Goliath and Red Lady is one of the most common ways people learn about Skydog and our mission.

Goliath's family spent a few years roaming on a twelve-hundred-acre section of the ranch, until Goliath started losing weight. We brought them back in to the biggest pen we have—the fifty-acre one where Goliath and Red Lady were first reunited—so we could better supplement his feed and keep an eye on his health. And today Goliath is still educating others and helping raise awareness.

Goliath and his family are great representatives of the rare curly-haired trait that naturally occurs in some of the wild horse herds in the western United States. His band is from the Salt Wells Creek herd in Wyoming, which has some of the most incredible examples of curly-haired wild horses who are also huge in bone structure and stature. We hope more people join us in the fight to prevent the complete eradication of this historical herd and the loss of the wild curly genes forever.

*Domestic horses demand constant care,
hoof trimming, shoeing, tooth filing,
immunizations, worm medicine, mineral
supplements. The only care wild horses get
is natural selection. Parents are not chosen
by stud books, but by the blows of competing
stallions. The desert prunes any deficiencies.
Wild horses may not look like much,
but in many ways, they are the best horses.
The wild has given them no other choice.
What emerged are animals that, according
to their riders, have unparalleled intelligence,
stamina, and overall resilience.*

—

David Philipps

*Wild Horse Country: The History, Myth,
and Future of the Mustang*

chapter

MAESTRO, GRACIE, LEGACY, AND BO

8

8

A couple of months after we adopted Goliath and reunited him with Red Lady, we were contacted by another wild horse organization who had taken in a gelding named Maestro who was way too much "big medicine" for them and was not settling well into captivity. They asked if he could come to Skydog. We had seen photos of Maestro with Goliath in the BLM corrals during the time they were captive, sticking close together and keeping each other calm, so we agreed.

When Maestro first came to us, we put him in the pen next to Goliath. But with eight-foot elk fencing between them, they sparred and fought, as Goliath had become fiercely protective of Red Lady and his newborn baby Bodhi. Obviously, whatever truce the two had reached in the BLM corrals was not carrying over into a situation where mares were involved. So we moved Maestro, along with a mare named Gypsy Rose, down a pen, and peace was restored.

Maestro was a wild one, wild and strong, and not interested in having anything to do with people. He would hide in the trees and absolutely refuse to come close, even for food. He seemed unhappy and still very upset and disturbed from his capture.

A few months had passed when I received an extraordinary email from a woman who had adopted a mare named Gracie from the same BLM corrals as Maestro. It seemed the mare had been his lead mare in the wild.

Gracie had recently had a foal named Maestro's Legacy—a black colt with a blaze. The woman had seen what we had done for Goliath and wanted Maestro to get his mare back, too.

We said yes.

It was incredible that another adopter cared enough about her horse's happiness that she wanted the chance to reunite mare and stallion. And so, Gracie and Legacy traveled to Oregon from Colorado. When we put them in with Maestro, a beautiful dance began. Maestro ran to them immediately, and Gracie greeted him, but she went to great lengths to keep their son away from his father at first. It was only a couple of hours later that the three of them were grazing together peacefully. Another family back together. Another wrong righted.

We thought that was that, until three years later, I received an email from a woman in Arizona who had a gelding named Bo who was believed to be from Maestro's band and likely Gracie's son. Bo had suffered an injury to his back and could no longer be

ridden. He had been receiving regular chiropractic adjustments and stretching to keep him pain-free, but his owner felt he was bored and unhappy. We said yes, we would take him—after all, he was part of Maestro and Gracie's family.

Bo traveled to our Malibu facility, and from there to Oregon. He is the most remarkable horse and stands out in a herd due to his extraordinary white blaze and orange curls. We wondered how it would go to introduce a grown son to a family with a possessive and territorial patriarch, who already had one son to contend with.

skydogsanctuary

Really there are two Maestro's for us—the one before Gracie and Legacy came to Skydog, and the Maestro after. We knew him as a shy, nervous, suspicious, and wary horse, who rarely came out of the woods and would only eat after we dropped the hay and left. We would catch glimpses of him, but he would never come close or relax and kept his guard up worried about people and their intentions. But then Gracie arrived and he transformed. To the point where he would lie down and roll right next to us showing he had not a care in the world that we were close. He has his Gracie back and their beautiful son and all the rest is behind them and hopefully now they trust this is forever and a life they treasure together. Maestro is one of the families we have reunited after roundups tore them apart and every one of them is so emotional. Their bonds, their love, their recognition and happiness at seeing each other again, no matter how long it's been. There might not be physical healing with these horses, other than the ones coming out of the corrals, but the psychological healing is just as important and evident so we will keep reuniting beloved family members, because family matters, freedom matters, and Mustangs matter  

MAESTRO, GRACIE, LEGACY, AND BO

← Legacy
↙ Maestro
↙ Gracie

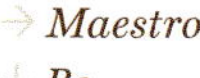

We turned Bo loose in the aisle next to Maestro's pen to gauge their reactions. Legacy came to say hello, but Maestro was not having it and charged the fence a few times to "protect" his mare. Gracie, Bo's mother, ignored him and did not even look in his direction.

We felt the verdict was in. Maestro had made it clear he wasn't interested in having another son back in his family, and we were fine with that. We were happy to add Bo to the "boys band" on the ranch, where he still runs today, part of the herd and looking magnificent. We say he's "flown the nest," as his family is nearby, but he has his own life elsewhere. He hasn't shown any signs of pain from his back, and we believe running the hills and moving constantly has helped heal him.

The best part of this story is that we gave Bo and Maestro and Gracie and Legacy a choice, and that's one of the most important things we do. For these horses who have had little to no choice in their captive lives, we allow them to choose who they like and dislike, where they want to be on the land, and how they want to live in conjunction with humans. We are so proud of that.

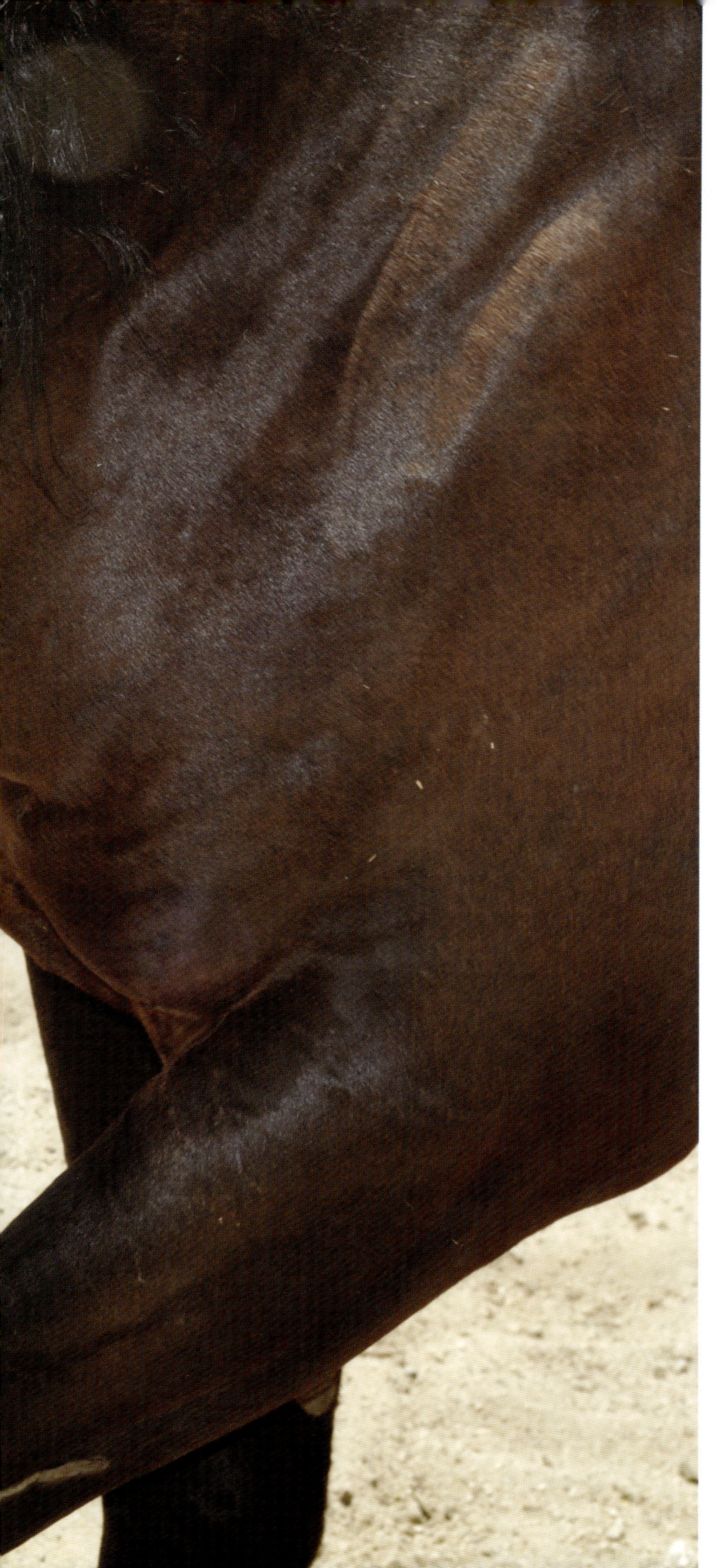

In 1971, Congress passed a law to protect wild horses. Since the BLM oversees so much unwanted land, and most wild horses live on the same unwanted land, it has fallen to the agency to oversee the horses too. When I say, 'oversee,' I mostly mean 'remove.' In an attempt to keep the horse population stable, the agency rounds up thousands of horses a year using helicopters that sweep across the desert and chase the herds into corrals, where they are trucked away and put in storage.

—

David Philipps

Wild Horse Country: The History, Myth, and Future of the Mustang

chapter

MAGGIE AND HER FOUR ORPHANS

9

9

Right after we rescued Goliath, I was on a mission to try to find any of his other mares. Goliath had a gray mare and a black mare, as well as Red Lady, and as all the mares from that gather had been taken to an off-range facility in Idaho, I registered to tour the facility to try and identify any of the family. It was a public tour, and we were all loaded onto a flatbed with hay bales to sit on and dragged by tractor around to view the pens.

I took hundreds of photos of any gray or black mares I could see, hoping to zoom in and identify them later on my computer. It was a sad and miserable sight—all these incredible horses packed into small corrals, especially as most mares had babies by their sides, and the pens were truly no place for newborns. Not only were they unsafe due to crowding, but there were no hills or streams to explore, nor was there room to run with playmates. Just barren filthy pens full of other miserable, bored, dull-eyed horses, just existing.

Once I had established that none of the mares I was hoping to find were at the facility, I changed my focus to posting photos of some of the mares and babies with the promise that I would help facilitate any adoptions out of these off-range corrals.

Quite a few people reached out about one particular mare and baby. The mare was a cremello with ice-blue eyes with a little grulla colt at her side. I had one woman in particular who wanted to adopt them both, and I contacted the BLM in Bruneau, Idaho, to ask if the pair could be transferred to the Boise corrals for adoption. The BLM agreed, but shortly after, they informed me that there was a virus at the corrals. Then another. Then a third. All summer long, I kept asking to move the mare and foal before they got sick, but I was denied every time.

Finally, I got the email I had been dreading—the grulla colt had died. Without the colt, the adopter backed out, so we decided to bring the cremello mare to Skydog to help her heal. It was all we could do after the ordeal she had been through. We drove to Boise, Idaho, to pick her up and traveled back to Oregon with just one mare, who I named Magnolia (Maggie, for short) in the trailer.

Rewind to a few weeks earlier: The corrals near Burns, Oregon, had asked if we could help some orphan foals. Some had lost their mothers or been found alone on the range during the roundup. One was rejected by her mother due to the stress and panic the roundup caused. Roundups are by no means the "nonlethal solution" the BLM portrays them to be. Many horses are euthanized at the trap sight for injuries or die at the corrals in the months following the

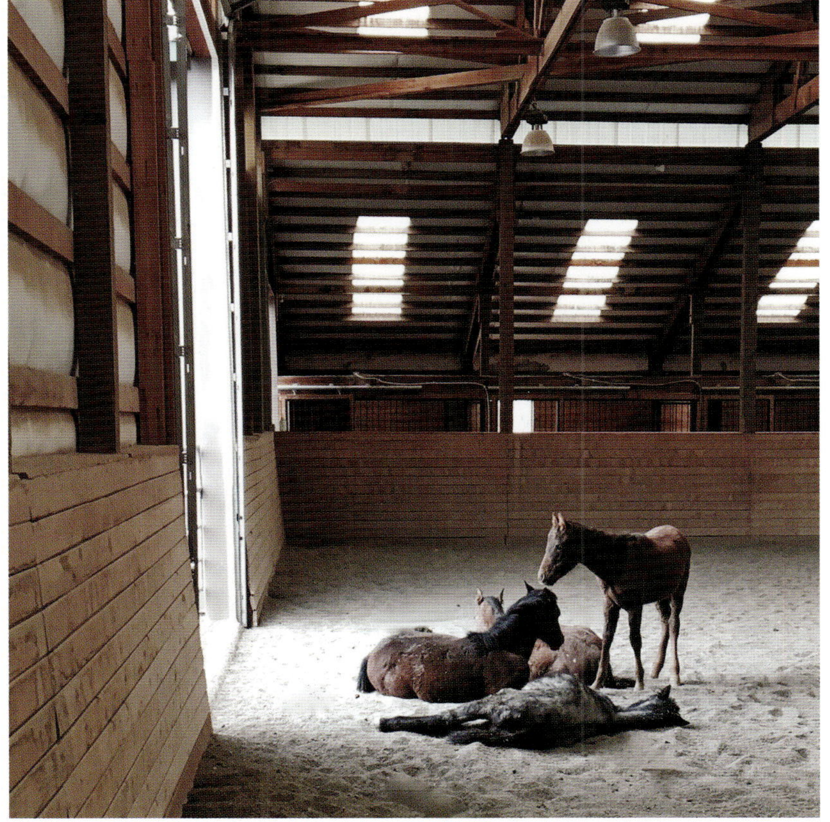

roundup as they break legs, necks, and hips being run through chutes to be processed. Unfortunately, since mares and foals are often part of the gather, orphaned babies can be the result.

We went to pick up four orphaned foals, and they were a sorry bunch—heads hanging low, no spark in their eyes, injured, and

skydogsanctuary

These horses matter. These horses embody our history in the West and need to be preserved and protected. Respected and admired not treated like feral, worthless, inconvenient invaders. They are smart, sentient beings who treasure their families above all else and we see and appreciate their value and worth in our ecosystem. This family represents all of that and more in terms of compassion and kindness being more valuable than greed and the accumulation of wealth off public lands. Our public lands, which are being used for grazing and plundered and pillaged for mining. So breathe this in because I don't want sanctuaries like ours to be the only place to see wild horses like this. We have to keep looking for solutions for our wild horses to live free and we must keep fighting for them. I vow we will never stop 💙

Magnolia (Maggie) with Coco, Two Spots, and Sugar

skinny. As soon as we had them home, we started on their rehab and care. The one we called Willow had a fractured jaw with pieces of bone that needed to be picked out. Coco was malnourished and had serious lacerations to the inside of his back legs. Sugar was tiny, frail, weak, and covered in wounds. Two Spots was healthy but was only a couple of months old—far too young to survive without a mother.

The team at Skydog fixed them up, caring for their wounds, cold-hosing their swollen legs, and giving them all the love they were missing from their mothers. The four orphans clung to each other for dear life and were very bonded and close. Gradually, they began to get their spirits back, and their health…and then all they needed was a mother figure.

When we got Magnolia, she was sad, too, having lost her baby, as well as her family and freedom. I had an idea:

What if?

Could we?

Would she?

We led Maggie down the hill to the barn where the orphan babies were huddled together, opened the gate, and led her inside. The four all came toward her, tentatively and curiously, one step at a time. They said hello. And then Maggie knew exactly and instinctively what to do next.

She gently, then more firmly, started grooming them. First Coco, then Two Spots, then Sugar and Willow. She did what mother horses do to soothe their babies, nibbling away at their necks and backs.

And they returned the favor.

*Like the bald eagle, the wild horse
was in pretty dire straits by the middle
of the twentieth century. They had been
hunted, killed, poisoned, rounded up,
and sold for slaughter until there were only
about seventeen thousand left. Then in 1971
Congress passed the Wild Free-Roaming
Horses and Burros Act, declaring the wild
horses 'are living symbols of the historic
and pioneer spirit of the West' and
'shall be protected from capture, branding,
harassment, or death.' President Richard
Nixon paraphrased Henry David Thoreau
at its signing, saying, 'We need the tonic
of wildness.' He said wild horses are a living
link with the days of the conquistadors,
through the heroic times of the Western
Indians and the pioneers, to our own day
when the tonic of wildness seems all too scarce.*

—

David Philipps

*Wild Horse Country: The History, Myth,
and Future of the Mustang*

chapter

10

The staff at the BLM corrals in Burns, Oregon, reached out about a mare who had given birth to twins. It was an absolute miracle, as only one in ten thousand mares conceives twins, and few take them to full term and give birth without a lot of help and veterinary care. This mare had been chased by a helicopter in the summer heat, processed, branded, and corraled in a filthy pen with other mares and still foaled out, without any intervention.

The BLM corral staff told me that one of the twins wasn't doing that great and she felt the foal needed extra help. Would Skydog take them? I was of course ready to do whatever we could to help the mare I called Elsa and her miracle babies. After all she had been through, she deserved extra nutrients, good feed and minerals, and most of all, a stress-free environment to bring up her twins in peace.

I went to the corrals, and the three had been separated from the other horses. I took a series of photographs, talked to them, told Mama Elsa what a good mare she was and how they were going to stay together and live out their lives in peace, with all the love and care Skydog could give them.

Elsa listened intently as she ate, while her two blue roan babies played around her. I passed the adoption by the board of directors and made plans to go and get them.

And then it all went sideways.

Suddenly, the head of the holding facility contacted me to tell me I would *not* be able to take Elsa and her babies. Elsa had been chosen to be part of a spay study—her ovaries were to be removed and then she would be released back to the wild. Her babies would be put up for sale, to the highest bidder, separately, with no mention of them being twins on the adoption profiles.

I couldn't stand the thought of putting the mare through more trauma, or separating the twins, who were bonded and very close. I started lobbying the corrals to use another mare for the study and let Elsa and her babies come to Skydog, as we had planned. They had, after all, *asked* Skydog to take them! We were an approved facility that BLM representatives had visited numerous times. There was no reason to cancel this particular plan for no good reason.

I wanted to keep Elsa and her twins together very badly. I believe in family and protecting and defending family where I can. And I think it is inhumane and arrogant to think humans are the only species who care deeply for their young.

So began Skydog's campaign to #freethetwins, along with a petition. We asked people to write and call the BLM to stick to their word and release all three horses to us. It all fell on deaf ears. In addition, the holding facility became hostile and uncooperative, indicating that going forward, we would no longer be allowed to take special needs horses, and they would be euthanized instead.

Months went by, and the online adoption came around. By now, the spay study had been canceled due to lawsuits filed by advocates. So, incredibly, there they were—Elsa, and the twins

skydogsanctuary

What a glorious experience to sit with these souls and just be. To have them check me out and for Barkley to stand gently hoping to make friends. He stays so quiet and still so as not to frighten them and has made friends with so many baby horses in his time at Skydog. The babies are more curious and friendly with every day that goes by while mama Elsa keeps her distance and a watchful eye. Feeling super blessed to be able to witness these moments and they feel like God's reward for all the work that makes these moments possible. Next week we have three more beauties arriving and I can't wait to meet them. Thank you to everyone who supports us and loves these horses as much as we do. So grateful today 💙 💙

Elsa

I called Hope and Promise. They were posted for the auction separately, with no mention of their relationship to each other. I bid on Elsa, and two incredible supporters from other states bid on a baby each in order to help reunite them with their mother.

The horses were picked up and brought to Skydog one at a time, and we joyfully reunited them. It was a glorious moment to deliver on what had been promised to this incredible mother months before. She had her babies back, and they had their mama. Sadly, the twins had already been weaned, which was heartbreaking, as they were still so young and would have been better off with their mother's milk. But, they had her and each other, and that was what mattered most.

↑ Hope and Promise

→ *Hope and Promise*

I spent weeks just watching, enjoying the babies' reconnection with their loving mother. I was under an enormous amount of stress at the time, as Skydog was now under constant attack, all for trying to keep a promise to a remarkable Mustang and ultimately succeeding in keeping a family together. Make no mistake, there is a remarkable hatred for rescues and sanctuaries within certain circles, and there are some who make it their mission to spread false information and lies, regardless of all the evidence to the contrary.

The BLM was furious for seemingly having lost a battle that never needed to happen. Eventually, they made peace with me, and I put the whole thing behind me. But I will never forget the unnecessary injustice done to these horses for no other reason than a desire for control.

Elsa and the twins still live happily together on the ranch. The twins are extraordinary to see—they change color from the lightest gray in summer to the blackest of black in winter. And of course they are loved regardless of what color they are that season and live a peaceful life—with family and each other.

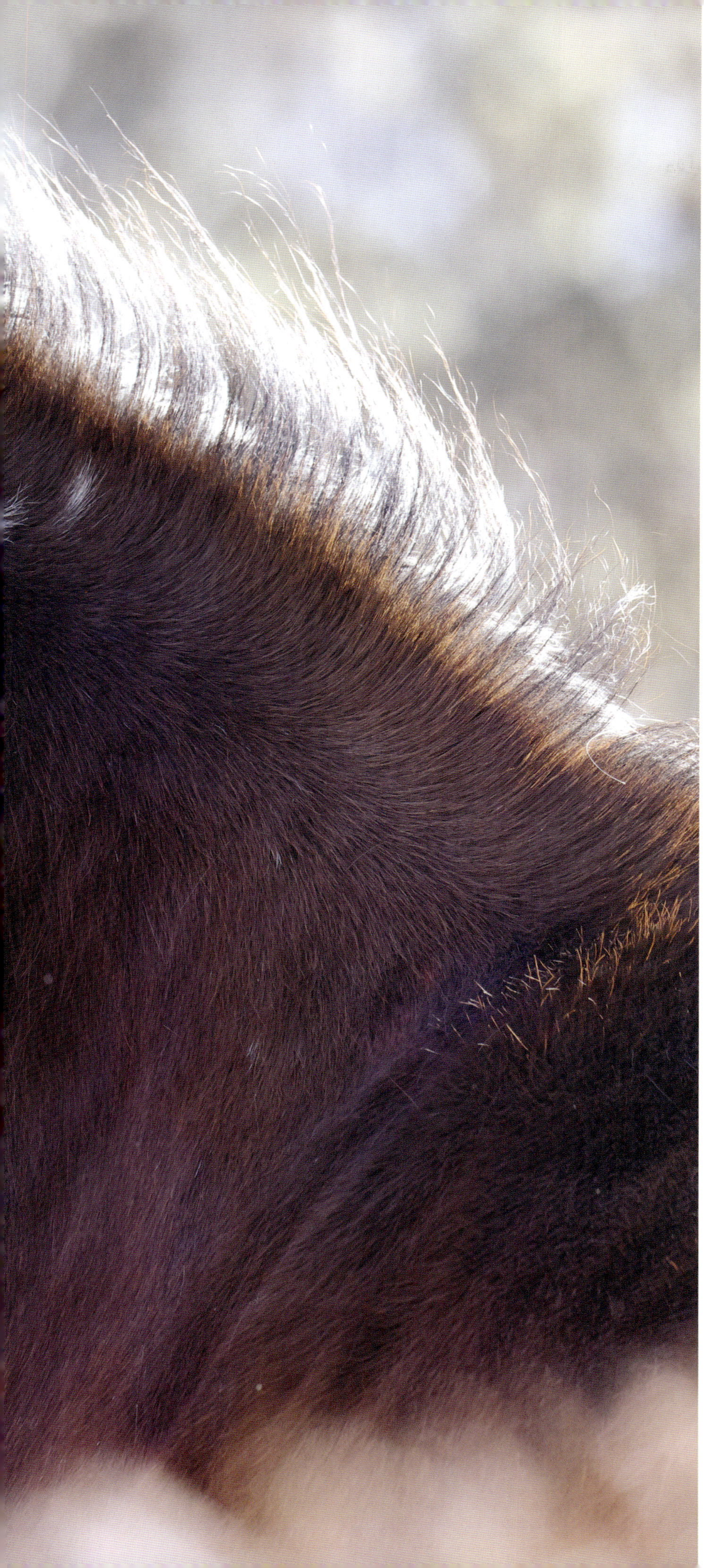

*It's clear that the horse's ability to provide
flight was universally desired, and nowhere
is this desire more pronounced, more
extreme, than in America, where escape and
the chance to start over is not a pipe dream,
but a birthright. We may not think
of ourselves as part of a horse culture,
like the nomads of Mongolia, for instance,
but in our own way, we are: we worship
cowboys, and we're jacked on freedom,
and we love moving forward through wide
open space, preferably on a cactus-lined
highway in our most iconic car, the Mustang,
whose grille features a galloping pony.
Yet as we ply the road, many of us do not
realize that the real thing is fighting
for its life on the rocky playas,
just over yonder, staking out the dream,
being wild and free for the rest of us.*

—

Deanne Stillman

*Mustang: The Saga of the Wild Horse
in the American West*

chapter

BLAZE AND FAMILY

11

Blaze was a horse I did not know of until I received a tearful call from Laura Leigh of Wild Horse Education about a Mustang family she had followed and admired for years. They lived on Fish Creek on a "herd management area" in Nevada that had seen a number of battles and controversy for years. We had just adopted Elsa and the twins, and I was still reeling from the hatred and hostility from a number of online groups who, for no apparent reason, had been infuriated by our work to reunite Elsa with her babies. I was honestly questioning whether I could go on doing the work of Skydog, with so much negativity and attacks that were vicious and uncalled for. It was affecting my health, and I don't believe I have ever spoken to another rescue or sanctuary who hasn't gone through the same thing.

Well, Laura asked if we could take Blaze and his family. I contacted the corrals that held them and they agreed—then for a couple of weeks I braced myself and tried to get my mental strength back to go through the possible abuse once again.

By the time I sent the adoption applications, I was told the mares had been shipped to long-term holding in Nebraska, and most of the rest of the herd had been sent to other corrals where there was no access. The only horse that remained was Blaze, and he too had been marked for long-term and was in a pen with others waiting to be shipped. I begged. I literally pleaded with the corrals to sort him so I could come get him, but they said they were busy and didn't have enough help to do it right before Christmas.

I was heartbroken and devastated to know that the hatred we had received, and the effect it had on me, had resulted in failure for this Mustang family. I would not be able to reunite them as I had hoped.

Then one of the corral managers called and said that he would sort Blaze, and I could come and get him in the New Year. I was so thankful and happy—it was the best Christmas gift anyone ever gave me. I decided then to get back up off the floor and continue to fight for these horses.

I then made a terrible mistake: I posted photographs of Blaze's family at the corrals in the hopes that people would watch their local BLM facilities and help us find them. I should have known better. The online abuse began again, and several individuals made it their mission to adopt Blaze's children to make sure we didn't get them.

First, they went after his son in an online adoption, and bid up to a crazy amount of money, thinking they were bidding against us. During that time, Skydog was mercilessly attacked for "bidding

against adopters," even though we had decided *against* bidding on this gelding as we didn't particularly think Blaze would appreciate a grown son coming home. We decided we would only bid if the horse had no offers. As I watched the number rise daily, with everyone assuming we were involved in the bidding, it amazed me how much spite there was toward our simple mission to reunite a family of wild horses.

When the auction was over, and it was finally realized it wasn't us bidding, the adopter apologized to us and offered the horse to

Blaze

Belle Star has found her feet and is trotting, running, and keeping up with her mama. Blaze is always close by watching and protecting—he is such a good father. This little baby is so special and precious. She is a divine little being and I am so happy she gets to grow up with her family. Thank you for believing in us and supporting the family reunions we have done. They are for sure the most challenging as they seem to bring out the worst in some people, which is mind-boggling to me. Thankfully none of these horses know any of what swirls around each one of these family reunions, but from Goliath to the twins to Blaze, each one has caused fury from groups on social media. It seems weird to me that peace and respect for these families has to be so hard-fought, both on and off the range. And fighting for this peace seems so contrary to what we are trying to achieve. But I would do it all again in a heartbeat and probably will one day. Blaze, Hannah, and Belle Star, just you keep on running and loving and we will keep on making sure you're together forever 🐎🐎🐎

↙ *Maia and Belle Star*
↙ *Blaze and Hannah*

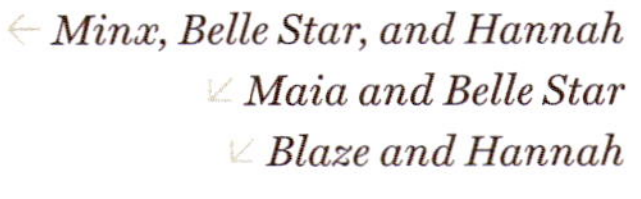

the second bidder, which was, thankfully, a wonderful home with a great trainer. Of course, Blaze's daughter was about to be offered in an online adoption, and I hoped that our detractors might not notice her, but they did. We never bid. But a different and wonderful sanctuary lost out at the last minute to an individual who continues to attack Skydog to this day. I hope the mare is happy and wasn't sold on once the glory of "winning" her was overwhelmed by resentment for having spent thousands of dollars for a wild horse who had nothing to recommend her other than her relationship to a particular family—a family who I am sure still misses her.

All this aside, there was a happy ending for Blaze. A month after we adopted him, we were informed that one of his mares from the wild had gone to an off-range facility in Bruneau, Idaho. I reached out and they agreed to allow us to adopt her. So two months after we brought Blaze home alone, Blaze saw his mare Hannah again for the first time. He came crashing through the woods to join her, which we caught on video! It totally made up for all the nastiness and politics, and reminded us that, in the end, repairing those broken horse family bonds was all that mattered.

*Most famous in our western annals and
Indian traditions is that of the White Steed
of the Prairies; a magnificent milk-white
charger, large-eyed, small-headed,
bluff-chested, and with the dignity
of a thousand monarchs in his lofty,
over-scorning carriage. He was the elected
Xerxes of vast herds of wild horses,
whose pasture in those days were only
fenced by the Rocky Mountains
and the Alleghenies.... The flashing cascade
of his mane, the curving comet of his tail,
invested him with housings more resplendent
than gold- and silver-beaters
could have furnished him.*

—

Herman Melville

Moby Dick

chapter

PHOENIX AND GHOST

12

12

IT is not often that a single photograph leads to a rescue and reunion, but that is the story of Phoenix and Ghost. I was looking at the American Wild Horse Campaign (AWHC) website one day, during the Warm Springs, Oregon, roundup, and saw a photograph of an appaloosa stallion trying to jump out of the trap he had been driven into with his family.

I stared at this photograph, taken by AWHC staff member Steve Paige, and thought how much it represented the freedom these horses want so desperately to keep. I had to know if this horse had survived his unbelievable leap toward freedom…or if he had lost his life trying.

I drove the ninety minutes up to the BLM Burns District corrals in Hines, Oregon, to search each pen to see if I could find him. With my long camera lens, I scanned each pen, filled with over a hundred horses, zooming in on any horse I thought could be the one. Finally, I found him. He was standing close to another horse I remembered from the photo; I actually saw his friend fight off any other stallion that came close to them and thought perhaps he was a "lieutenant" from the herd.

I also searched the mare pens and there, milling amongst dozens of other horses, was the white appaloosa mare I had also seen in the photo, looking up at her stallion as he struggled to scale the immense trap fence.

I named the stallion Phoenix that day, and I was told by the BLM staff that he and his friend had been selected for return to the wild. My heart was happy to hear that he hadn't lost the freedom he had jumped so high to keep.

I regularly visit the Burns corrals, and every time I went for the next twelve months or so, I looked for Phoenix and his buddy. I always found them together and would confirm with the staff that they would soon be returned to the wild. I never saw his white mare there again, although I would always look.

One day I saw Phoenix looking lost and alone in a different pen. I asked the head of the corrals what had happened and was told they had cancelled his release as they had done an aerial search and found enough horses in the Warm Springs HMA. Just a number, without any regard to the horse they had held for months with the promise and intent to release him.

So, I put in the adoption paperwork for Phoenix because I knew I had to try and give back to him what he had lost—his freedom, his mare, his babies, his best friend.

It was a wonderful day when Phoenix arrived at Skydog and leaped out of the trailer and back to the wild life he had wanted so badly. As time went by, he settled in. First, we put him with Blaze, who at the time was also waiting for his family to be found. The pair spent months together, keeping each other going. Then Blaze got Hannah back, so Phoenix was put with a group of friendly geldings in a bachelor band, and he seemed content. Still, I noticed he would often stand alone, staring off into the distance, as if he was watching for something, scanning the horizon in the hopes of seeing his white mare and family. Whenever I visited the BLM corrals, I scoured the pens for his mare. I always came away feeling sad not to have found her.

One day I was going through the BLM online adoption profiles, and I spotted a white mare and stopped scrolling. I looked at her, and then I looked more closely, then went to the video they had shared with potential adopters. There was no mistaking Ghost, as I'd come to call her. It was the white appaloosa mare I had been searching for, running back and forth in fear in the video. Her features matched in every way, and right at that second, I placed a bid. Her freedom was won for a thousand dollars.

→ *Phoenix with his best friend at the BLM corrals*
↘ *Ghost after being run too hard into the trailer at the BLM corrals*
↘ *Ghost*

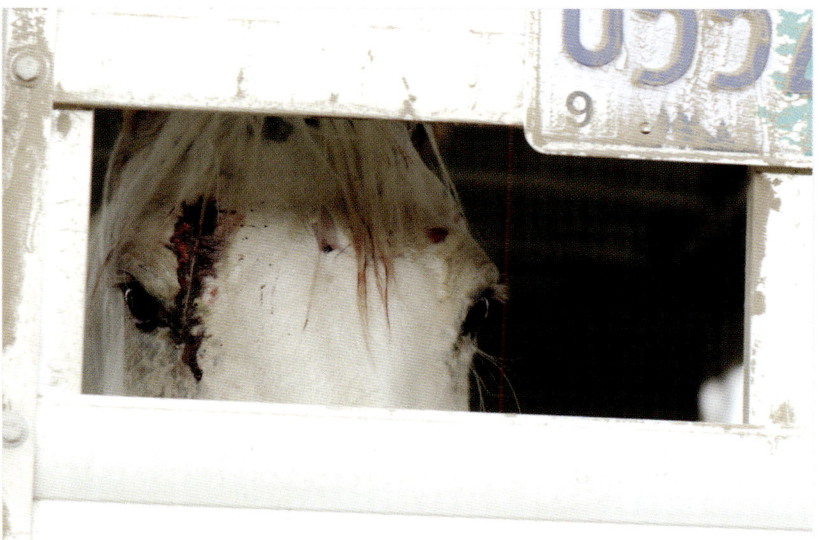

skydogsanctuary

Talk about making an entrance. In a cloud of dust, sunlight glinting off her pale coat, Ghost bursts into Phoenix's pen as if she knows she is running toward her destiny. Phoenix sees her and calls out to her and she hears his call. High stepping, tail flagging, head high as she can raise it, she dances toward him. Way up on the hill above, Phoenix stands staring as if he is seeing a ghost. He looks and makes sure it's really her and starts to run toward her, tripping as he goes. Closer and closer they get. And then the meeting—the turn of their heads like two swans, the deep breathing each other in, nose to nose and cheek to cheek they stand. Always in the wild it would be Ghost who led and Phoenix who followed. They go to move, see something, and take off together side by side, off to a quieter spot to reacquaint themselves. *Well where have you been and how did you get here?* And then they both gaze out at the views surrounding them, taking it all in before heading off down the hill to explore. And just like that, wrong is right and love wins the day. They are together again and suddenly for a moment everything is okay because two wild horses found their way back home to each other. Never to be separated again. Forever and ever. Amen.

→ *Phoenix and Ghost*
↓ *Phoenix*

Sadly, the BLM wranglers ran Ghost so hard into the trailer she smashed up her face. Her eyes were bloodied and swollen by the time we got her to Skydog. It was heartbreaking to see. We put her in a pen with two friendly mares to let her heal physically and emotionally from all she had endured.

Finally, the day came when she was well enough and strong enough to be released with Phoenix. Ghost and her new best friend Blue Moon, who was also an appaloosa from Warm Springs, ran into the pen with Phoenix. It was a special and spectacular moment when he saw her again, greeted her, and ran off up the hill with her to reunite—for forever now.

Phoenix was so proud and visibly different in every way from that day forward. He walked with his head high and his entire body seemed to relax. His whole demeanor was changed by having his mare back. Blue Moon stayed with him and Ghost until she went too blind to be able to cope in such a large, open space.

Ghost and Phoenix continue their grand love affair at Skydog. And Phoenix, the horse who had sacrificed his body in a dramatic leap for freedom, seems content. Because he has the one thing that matters most—family.

The legend of the wild horse—all that stuff about freedom and toughness, which secured its place as an American icon? It is well deserved. Like nearly all Americans, the wild horse is an immigrant. And like many it prospered through sheer grit. The herds on the land now are the descendants of the painted war ponies that allowed a few thousand native warriors to hold off the industrialized American army. They're also the descendants of the cavalry mounts that chased down Crazy Horse and Geronimo. They are the descendants of the Pony Express runners that whisked messages from the Mississippi to San Francisco in ten days before the invention of the telegraph, and the cowboys' tireless sidekicks in the great cattle drives.

—

David Philipps

Wild Horse Country: The History, Myth, and Future of the Mustang

THE PINE NUTS

13

13

The "Pine Nuts" are a herd of horses living wild near Gardnerville, Nevada, with an international following. Theirs truly is a model of good management of wild horses: advocates not only follow and photograph and name the horses, sharing their stories online, but also go out on the range and dart the mares with birth control to help control herd numbers. Sadly, human residents in the area sometimes complain about the wild horses coming onto their unfenced property, which leads to many being bait-trapped and removed. This is heartbreaking for the thousands of people worldwide who have grown to love this herd, watching new babies be born, stallions fighting over mares, and other aspects of wild horse life, all in real time.

A couple days before Thanksgiving in 2020, the BLM trapped an entire Pine Nut family, one that was especially loved and followed. It included Samson, the red roan son of the most famous herd stallion Blue, and at that time the leader of a group of mares that spanned four generations—from twenty-six-year-old Ol' Momma, to her daughter Apple, her daughter Dumplin', and her baby boy Sam. Also captured with this group was Jet, Samson's brother, who lost his own mare and daughter who remained outside the trap when that gate clanged shut.

↓ *Samson sparring with his father in the wild*

The band was slotted to be auctioned off online and separated forever, unless someone stepped up to keep them all together.

And we did.

The entire Pine Nut family was still together when I traveled to the Palomino Valley Wild Horse and Burro Center near Reno, Nevada, to pick up Blaze. Knowing their story, I walked over to see them. It was extraordinary to see this incredible and famous wild horse family there in that filthy pen, the stallions still swollen and in pain after their gelding. They were separated from their mares, but I was happy to see baby Sam had been left with his mother and not weaned. Some small mercy.

And so the mission began to bid on each of them. At the end of the online adoption period, as I clicked from horse profile to horse profile, seeing the notification that our bids had won had never looked better. The entire family was safe and coming to Skydog. What a win indeed.

And so we traveled back to Palomino Valley and picked up the entire family (along with a donkey named Dora I had promised to come back for). It was like a military operation: we took two large trailers and a massive crew and drove home in a convoy with our precious cargo. As the sun set that evening, we unloaded them all and breathed a huge sigh of relief. The Pine Nuts were home.

And that was when their new lives began. Sam was still tiny, so for a few months, as we do with all new babies, we kept them in a smaller pen together, just putting on weight, Samson and Jet occasionally sparring, and the family dynamics ever-changing. The mares had remained so close through it all—Ol' Momma doting on her new great-grandson, helping her daughters in raising and protecting him. She would steer the whole family away when we came in with food; they kept their distance, thanks to Ol'

Momma's insistence. Sam was the only one to venture close out of curiosity and rebellion, only to run home to his mama and aunties if we stretched out a hand in his direction. He was a perfect little colt, and full of beans and such a scamp. His family doted on him and let him have pretty much free rein.

When at last we could turn the Pine Nuts out on a few hundred acres of their own, we noticed that more and more, Jet was alone or standing at a distance from the group, as Samson seemed to become more protective and possessive of his mares once they had more space. It made me sad to see him like that. I remembered

how he'd lost his mare and their baby in the roundup, and I felt a sad longing in Jet's heart for them.

During the next online adoption, I saw a photograph of a mare from the same herd, from Pine Nut Mountain, and in her photograph she was sticking her tongue out at the camera—and those who had torn her from her family and life in the wild, it seemed. She had zero bids, so I bid, and when the clock ran down, the mare was ours. Or rather, Jet's, as the plan was to add one more mare in need to his family and see if our "dating services" were appreciated and Jet would find love. His own love, just for him.

Thankfully, it worked. Jet and Mystery became inseparable, and little by little the other mares accepted her, too, and they became one big happy family. Young Sam also took a great deal of liking to Mystery, and the two are often off on their own together with Jet, just the three of them, with Samson keeping the rest of his group close.

There was a big debate about whether baby Sam was Samson's or Jet's son, so we tested his DNA. He turned out to be neither's! We hope he is the son of a stallion who is still wild and free, running up there on Pine Nut Mountain. Forever.

Jet

The snow makes the whole ranch look incredibly beautiful, magical, and like a real-life fairy tale, with the horses blowing clouds of steam and the icicles in their manes making music. The horses love these colder temperatures much more than the heat of the summer. One of the things I think about these horses is how docile and gentle they are for large and strong animals. They have fight-or-flight instincts built into them but rarely choose fight with humans. They have such a kindness in them that to their detriment they willingly load onto trailers driving them to slaughter as they are so keen to please the human and do as they are asked. They take the punishment and cruelty because they don't have it in their nature to attack back or hurt people—or very, very rarely when provoked beyond all that's tolerable. Seeing these incredible beings, these gentle souls captured in such powerful bodies, is awe-inspiring and a huge testament to their quiet understanding of our shortcomings and emotional intelligence in their ability to forgive. They truly are the most forgiving souls I have ever met, and they inspire me to emulate that same quality and wisdom.

*I had come to the valley with the idea
that roundups were necessary and as humane
as possible. I had believed the agency
when it said wild horses are overpopulated.
I knew above all that the priority should be
to protect the long-term health of the desert.
But watching the families broken up
was wrenching, and it made me wonder
what kind of system we had created.*

—

David Philipps

*Wild Horse Country: The History, Myth,
and Future of the Mustang*

chapter

THE BALLERINAS

14

14

WE were contacted by the BLM office in Rock Springs, Wyoming, to see if we would consider taking two mares they had at the facility. They sent grainy photos of a couple of muddy horses, hiding together behind others, and said that many people had indicated interest in adopting the pinto, but none were willing to take the gray who was so closely bonded to her, as well.

We were told the two Mustangs were inseparable and even walked in sync. It was strongly believed they were mother and daughter. So we said we would take them at Skydog in order to preserve their bond.

The mares weren't ready to ship to Oregon for a few months, and I almost forgot about them until I received an email that indicated we could pick them up. A couple days later they stepped out of the trailer at the sanctuary, and all of us were shocked to see two absolutely gorgeous mares—one a beautiful palomino pinto (Autumn) and the other a sweet-looking gray (Swan) who, as we'd been told, stayed lock-step with her companion. It was extraordinary to watch them. I have never before, or since, seen such bonded and connected horses. It was as if they read each other's minds as they turned together, moved together, and ate together. They would both move their left foot at the same time, swish their tails in the same direction, and move in tandem as if they were doing an elaborate dance for the world to see they loved each other. They were inexorably linked in such a deep and beautiful way, we nicknamed them "The Ballerinas," and it stuck.

After we fed them back to health and they got stronger, we turned The Ballerinas out. They hid so well we weren't sure if they were even with us still! We would ride out on quads to look for them and often surprise them, together as always. They would stare at us for a few seconds, and then all we saw was a flash of tails and a cloud of dust. They wanted nothing to do with the two-legged creatures who had taken their freedom and wild home away, and we didn't blame them.

It took a good three months for The Ballerinas to join the herd of horses they shared land with at Skydog, but when they did they took on the roles of lead mares in no time at all. They were always the last to eat, the highest on the hill in the role of "look out," and doting over any babies and each other.

Autumn and Swan also became two of the most popular and most photographed horses we have at the sanctuary. Although

they keep their distance from people, they formed a tight new bond with the herd that has been wonderful to see.

In the summer of 2022, we spotted a palomino pinto mare in a kill pen in Kansas. I was immediately struck by her unbelievable likeness to Autumn. I took a photo of her brand and asked the BLM to read it and tell me where she was from. Sure enough, she was from the same small herd as The Ballerinas, on Green Mountain in Wyoming.

I am not sure there is a tonic better than being in nature with wild horses. There is a quote by author J. Frank Dobie: "No one who conceived him only as a potential servant to man can apprehend the Mustang. The true conceiver must be a lover of freedom—a person who yearns to extend freedom to all life." And that is how I feel about these wild horses. Of course, it is nice to have ones with us who are approachable and friendly, but there is nothing as beautiful as wild horses who take off running once they have spotted you. The flash of a turning horse, tail flying behind, the sound of their feet hitting the ground as they run over the hill. That's the way to truly love a wild horse. The beauty of wild is what we are all striving to protect. To admire these creatures from afar, respect their freedom and choices, to love them for being exactly what their description encompasses—wild horses.

I knew we had to save her, just in case she was related to Autumn and Swan. I just couldn't pass up the opportunity to possibly bring one of their children home. So she and another dun boy traveled to Oregon. We named the new mare Rising Sun as a tribute to Elvis Presley, as that was the name of his personal horse, a golden palomino he rode around Graceland during his happiest years. It's so good to have a third ballerina out there with the Skydog herd, shining her beauty and Wyoming Green Mountain genes to the world.

I often stop and think about all the horses at Skydog and the original photos that may have caught our attention or convinced us to take them. I truly believe each horse was meant to be with

↑ Adeline with Swan and Autumn

→ *Rising Sun with Swan and Autumn*

us and that there is some magic behind each story. We give the horses we save sanctuary for life, but the awareness each of them help raise for *all* wild horses, the education they help us give the world about the plight of the Mustangs, that is ultimately what is so important. I have watched as dozens of new sanctuaries have sprung up in the United States, doing the good work of saving wild horse families and keeping them together, and *that* is the culmination of all the work we have done. By showing the bonds the Skydog horses share, we have succeeded in convincing others in the world that these family ties are real and need to be respected.

The Ballerinas are now living wild and free on our nine-thousand-acre ranch. They have always kept their distance, and we are so glad for that. We took them in to *give them* their space, wildness, and freedom, and we honor that every day by asking nothing of them and leaving them in peace. They are a beautiful embodiment of the importance of allowing what's wild to remain on the range.

*There is no other animal in America that
we have heaped up with so much meaning.
Though the bald eagle is the country's official
symbol, it isn't as American as the Mustang.
The eagle is too regal and aloof—a symbol
of federal power, but not of American grit.
Besides, the bald eagle is known for stealing
fish from other birds, which prompted
Benjamin Franklin to call it a bird of
'bad moral character.' The Mustang,
on the other hand, embodies the core ideals
of America. It is not pedigreed. It has
no stature. Instead, it derives its nobility from
the simple toughness of its upbringing
in a free and open land. It is beholden
to no one. It will not be subjugated.
It is superior to his domestic brethren,
because it has the one thing Americans say
they yearn for most: freedom. It is the hoofed
version of Jeffersonian democracy.*

—

David Philipps

*Wild Horse Country: The History, Myth,
and Future of the Mustang*

chapter

BLUE MOON, DAKODA, AND LILY

15

15

Sometimes families aren't born families, but are forever bonded in some other beautiful and tender way that can count more than blood. The beautiful mare Blue Moon was one of a handful we were offered by the BLM Burns Wild Horse Corrals in Hines, Oregon, as senior horses who hadn't been successfully adopted. The staff told us that Blue Moon had always been seen with a friend who looked identical, but on the day their group was to be shipped to long-term holding, Blue Moon came up lame and was held back at the corral. They said she now seemed lost without her friend, and we agreed to take her to help give her some solace in freedom.

The moment Blue Moon stepped off the trailer, I knew something was wrong. The mare walked right into me, and it was obvious she couldn't see. I immediately felt even worse about the loss of her friend, who must have been her "eyes" and who she relied entirely on to be able to get around. I couldn't imagine how desolate she must feel without her friend or family member to help her. She bonded with Ghost (Phoenix's friend, who I told you about earlier in these pages) as she was in a pen, recovering from the wounds she had suffered from running into the back of the trailer while being loaded at the BLM. When the day came to reunite Phoenix and Ghost, Blue Moon barreled through the gate, too, unwilling to give her new friend up. We allowed

Blue Moon a few days out but soon realized it was too much for her with her lack of vision.

We brought Blue Moon back in to a smaller pen and put her with the mare Angel, who we knew at the sanctuary as an absolute miracle worker. Blue Moon had corneal ulcers in her eyes and needed daily treatment, and every day Angel would come through the chute with Blue Moon and stand right behind her while she was having her eyes treated. Angel would then exit with Blue Moon and lead her back to her pen.

We didn't want Angel to have to stay in a pen forever, so we introduced a third mare to the group. Her name was Dakoda. She was a mare that the BLM adoption staff in Cañon City, Colorado, had asked us to consider taking. She had been adopted out and had been well gentled and trained for riding, but the adopter had learned Dakoda had equine protozoal myeloencephalitis (EPM), a condition that was going to be very expensive to treat, and so he had asked the BLM if he could return her. I was told the man had cried when he said goodbye to the mare, so the corral manager had promised him she would find Dakoda a good home where she would be well-treated and cared for.

Dakoda needed daily treatment for the EPM as well as urine scald on her back legs, and so she seemed like a perfect addition to Blue Moon's pen so we could let Angel back out to enjoy her freedom. Dakoda and Blue Moon bonded easily, and we watched thankfully as Dakoda helped Blue Moon get around and was a constant and gentle friend.

And so these two special needs mares, both receiving regular medical treatment in the chute, remained in the pen by the elk barn. Little did we know that another poorly Mustang was coming to join them.

We were at the BLM corrals in Hines, Oregon, where we'd gone to pick up three donkeys, when I looked toward the medical barn to see the saddest-looking, sickest, most frail and unhappy filly, standing alone in a small pen. I took some photos and then went to the adoption office to ask what her story was.

I was told the corral management had planned to get a vet out to see the filly, as they truly didn't know what was wrong

When I think back on my life I can remember with such vivid clarity so many of my happiest moments were spent with horses. I have had this image in my mind lately of when I was in my twenties and living in New York City for work. I found a riding stables called Claremont, which had also been the name of the school I went to in England. It was a few blocks from Central Park and you could "rent a horse" for a couple of hours and ride around the sandy paths of that beautiful park. One spring day a man rode up to me and told me to stay still, and he pulled the branch of a blossom tree above me, and it showered me and my horse with pink blossom petals, and then he rode off. Isn't that a crazy, silly memory? But there are so many I had with horses that were the best times of my life.

with her. She had been slotted to leave for an adoption event but had turned so poorly, she'd been kept behind. The administration agreed to let us adopt her on the spot, and we took the lonely filly home that day.

Often the BLM corrals do not have the budget or time or inclination to treat seriously ill or injured horses, and these individuals may be euthanized by gunshot. I always feel that every "special needs" horse we are able to take is a blessing; it is truly the difference between life and death for many.

When we got the filly, who I called Lily, home, she was examined by our veterinarian and diagnosed with Pemphigus foliaceus (PF), an autoimmune condition where horses make antibodies

↑ Dakoda, Blue Moon, and Lily

→ *Lily*

that fight their own skin, potentially threatening their life. The disease had left Lily covered with scabs and clumps of falling-out hair, her belly and legs swollen, and anorexic and depressed. We began the necessary treatment to nurse her back to good health.

As she recovered, Lily blossomed, like a caterpillar into a butterfly. She was so beautiful it was hard to imagine she was the same sad filly we had adopted that day at the BLM facility. She had this perfect white mane, parted down the middle, curly as candy floss, which gave her the appearance of an "angel pony" and took everyone's breath away.

Because Lily still needs daily medication and treatment, when she was well enough, we added her to the pen near the chute with Blue Moon and Dakoda. The pretty filly settled in fine, with two doting aunties loving on her and protecting her, one on each side of her, day in and day out. And so, a "family" of three special needs wild horses was born and has remained together since. All three of them had their own, individual journey to Skydog, but now they are so bonded that I cannot call them anything but family. And what a beautiful trio they make. You would never know they were in any way disabled or vision-impaired.

Happy endings, one and all.

Their thoughts and memories are different from ours, but we also share some similarities. For example, Mustangs love their families, and they value independence. In the era when the wild horses of the West numbered in the millions, Mexican caballeros were hired to round them up for sale in stock in the Southwest. Then, as now, the act of taking Mustangs from their homes presented challenges. Horses injured themselves in the fight. In the process of being taken into custody, some died as a result of broken legs or necks. In some cases, however, the men hired to round up Mustangs describe stallions dying, not of physical injuries, but of broken hearts. The old cowboys called it Sentimientos. The stallions became distraught because they missed their mares.

—

Chad Hanson

In a Land of Awe: Finding Reverence in the Search for Wild Horses

chapter

RENEGADE AND FAMILY

16

16

This may well be the hardest, most painful story for me to tell. Because unlike most of the others I've shared, the ending is not what we had hoped for. Pope Francis declared 2021 the "Year of the Family," and with many people celebrating this idea, we decided we would try to rescue one large wild horse family and bring them back together.

A number of years earlier I had received an email from a young woman during a BLM online adoption event. She was bidding on a Mustang from the South Steens HMA, south of Frenchglen, Oregon. This particular horse, known as Renegade, had been photographed and videoed often in the wild, as South Steens is easily accessible to people, and it has a watering hole many of the horses frequently visit, offering a prime photography spot.

I had never followed this herd and didn't know the names of the horses, but I encouraged the woman to go for her dream horse, and I donated privately to her effort. I also said if giving him a home didn't work out, Renegade always had a spot at Skydog. I never thought she would take me up on the offer, years later.

By 2021, the young woman had also adopted Renegade's lead mare, who she called Lupine, in an attempt to make her dream horse happy, as he wasn't settling well into training and domestic life. Desperate to find a solution, she reached out, reminding me of the offer I'd by now forgotten I'd made all those years ago. But I of course honored the promise, and given the timing, we also decided that it would be Renegade's South Steens family that we would help reunite.

When our hauler, who has brought literally dozens of wild horses to us and other places, picked up Renegade and Lupine, she messaged me and told me, "This boy is special." It was the first hint I had of just how special he was. When he came off the trailer and into our arena, I sat in the sand and just stared at him. There was just something about him I couldn't put my finger on, and I couldn't take my eyes off him.

A few days after they arrived, we ran Lupine and then Renegade through the chute to worm them and trim their feet. But when we opened the door to halter him, we found a trembling, shaking, terrified horse, and I immediately said we should just open the gate and let him out. I didn't want to do anything to him at that moment, as I didn't want his first experience of us to be that traumatic. I had never seen a horse so petrified of being handled. I could see in just that one second how much this horse needed

to be wild again, far away from confinement and humans who came along with it.

Renegade's adopter knew the South Steen families well, and on her next visit to the BLM corrals, she identified Renegade and Lupine's daughter Dahlia, and her daughter Fern—and both had baby boys at their sides. Recognizing what an amazing miracle it was that we now had more family to reunite with Renegade and Lupine, we traveled to the corrals to adopt them.

While we were there, we noticed a very young foal struggling to walk and keep up with his mother. I went to the adoption office and asked if there was some way we could help him. I offered to take him to the nearest veterinary hospital. I offered to go into the pen myself to pick him up and support him, as he kept going down. The staff refused my offers and said they were monitoring the baby.

Two days later I received an email saying, "The baby died," with a sad emoji.

And so, when we went to pick up Dahlia and Fern, we added the dead foal's mother, Paisley, in an attempt to ease the pain of her loss. Home to Skydog we traveled with three mares and two babies.

Maple, Rebel, and Fern

What a moment 🧡💕 We turned Lupine's family out this week and I went looking for them. We had opened the gate for the big herd to come through to graze. I was thrilled to be there the moment the herds met. What a stunning sight to see them run, green grass below their feet and all the horses calling to each other. It never fails to make my heart soar when I think back to seeing these mares for the first time with tiny babies at their sides, born in dirty pens at the corrals, Paisley's baby dying, Cookie being returned, starved after adoption. All those images are in my mind juxtaposed with these images. So it means a lot. And how brave and fierce Cookie has become as the protector of this band. Wise and confident beyond his years, leading them toward their bright new future. And Lupine making my eyes fill with tears to see her with her family, enjoying her new life. Here they are and thank you to all who support us and them 🧡💕

← Cookie
↙ Renegade

Renegade and Lupine were so happy free on the couple of hundred acres we had turned them out on, we held off bringing Renegade in to reunite with his family. A thousand times I have regretted that decision, but I wanted him so badly to stay free and happy, to not feel restricted in any way.

One night before we planned to release the mares and foals with Renegade and Lupine, I went out to spend an hour or two in their company. I spoke to Renegade about his family being in the barn and safe, and I told him how they would join him soon. I left them as the sun was setting. Renegade was grazing on the grass along the river, and I remember turning back to get one more look at his beauty.

The next morning, I got a call over the radio saying there was an emergency in Renegade's pasture. I raced there and could see from a distance that he was down, and I thought he must be colicking. But as I reached him and ran to his side, the Mustang took his last breath and left this earth.

You have no idea how hard it is to recount this story, as even now, years later, it brings me to tears. I still have never found the right adjective to embody all that was Renegade. He was greater than words or sentences or poems or sonnets. I will forever miss him and feel an ache in my heart when people mention him. I loved him so deeply and am so forever sorry for everyone else who also loved him as a South Steens stallion and feels his loss.

Lupine stood guard over Renegade until the end, keeping coyotes at bay, until he left us. Then she took off up the hill and watched our human work from a distance. We brought in a kind gelding named Chief to tempt her back, and she slowly made her way down and then quietly followed him out of the pasture she had shared with Renegade, with one last look back at her love.

After Renegade's passing, we immediately reunited Lupine with her daughter and granddaughter, and their babies. And now they are a family. A family led by a matriarch who is fierce and strong, and at the same time, so dainty and wild. Lupine will still often stand at a distance from her family, as if remembering Renegade.

↑ *Renegade's family on the move*

→ *Fern with Renegade's family*

At some point we added Cookie, a young gelding who had been through a lot of trauma—he'd been adopted, starved, and then returned to the BLM in Utah. Somehow, along the way, he grew up, and when we turned the Renegade family band out with the big herd, Cookie stood tall, he and Lupine finding their wild selves again and taking on the roles of leaders.

Beloved Paisley, who had lost her tiny baby at the corrals, found a best friend in Maple, and we can only believe by the strength of their bond that they knew each other from their wild days or were related. Maple, who had been saved from euthanasia at the corrals, had been struggling to fit in at Skydog—until she found Paisley. They have been together ever since.

Renegade's family, and its adopted members, live as one and love each other fiercely. You will always spot them with the high white socks they inherited from their sire, the legend, and the most loved and missed horse ever to grace our land.

The day after Renegade passed, I looked up and saw the most glorious golden eagle soaring above the spot where I'd knelt beside him and heard his final breath. I remembered the myths and legends about horses and their unique ability shared with the gods to cross over into other dimensions effortlessly. In my mind, that is just what Renegade did. He wanted freedom without fences, and he found it, soaring above us in the skies, looking over his family from above.

Now, I continue to follow the Mustangs
as they wage their own last stand,
battling to save themselves in the high desert
where they went—like a lot of misfits—to be
left alone and flourish and finally
to hide from those who want to destroy them,
including not just those who killed them
in 1998, but a parade of such men
who have carried out acts even more horrific
as all the others who want the land
for livestock, fuels, and themselves.

—

Deanne Stillman

Mustang: The Saga of the Wild Horse
in the American West

chapter

HERA HERD

17

17

WE were finding a huge number of Mustangs in kill pens, and we were doing whatever we could, whenever we could, to help them out of there. The Adoption Incentive Program introduced by the BLM, which allotted up to a thousand dollars to qualified adopters, was resulting in dozens of people adopting wild horses, often as yearlings, keeping them for a year, and then dumping them in the slaughter pipeline. The rescues and sanctuaries simply cannot keep up with the enormous number of Mustangs landing in kill pens. We have, over and over again, explained to the BLM that this monetary incentive is just a "subsidy to slaughter" and a "pay to slay" program, which is hurting wild horses worse than anything else has thus far.

As we were gathering evidence about these kill pen horses to share with the BLM, we were notified of an injured bay mare who had an open draining wound on her back leg, which looked to be badly infected. We knew we had to help her before the infection got worse, so we jumped into action, bailing her from the kill pen and transporting her to a veterinarian who cleaned, treated, and dressed the wound. The mare, named Izzy, recovered at a quarantine in Kansas.

A month later, she traveled to Oregon where we continued her dressings and treatments in the barn. Before we turned her out with a herd, we remarked that her belly was getting remarkably round. Amazingly, in December 2020, Izzy delivered her baby in front of all the Skydog staff in her stall. It was a first for Skydog, as usually Mustang mares try to hide away and foal out with nobody around in the middle of the night. Witnessing a new life coming into the world was beautiful, and we were so glad that in rescuing one injured mare, we had also saved an unborn baby. We named the foal Wildheart.

At the same time we had brought Blue Zeus to Skydog, we were asked to take two mares from the same facility who had gathered in the same herd area—Hera and Strongheart.

Hera and Strongheart

skydogsanctuary

I rushed out to check on Hera when the sun came up and found her protecting her old friend, and it was as if the light had come back on for Strongheart. Her head was high and eyes bright—my heart just soared. This beautiful mare Hera who was never supposed to come here but somehow the angels had other plans, had healed her heart. So now they have each other and I know this will be the start of Strongheart's recovery back to her old self. She can talk about old times on the range and Hera will protect and guide her. Whenever I doubt, there is that nudge from the universe saying, "You're on the right path," and I am just so thrilled that they have each other in the best retirement we could give them. It's all meant to be and oh so beautiful. It's a glorious sight for sore eyes. Heaven sent ❤️💙🔥🔥❤️💙🔥

We had been told both had loved babies in the wild. Strongheart, in particular, had been seen taking in orphaned foals as her own and raising them alongside her own children, loving and caring for them deeply. When we rescued Strongheart, she was so starved and emaciated, we weren't sure she would recover. She also seemed cloaked in a great sadness. We had seen old photos of Strongheart with Hera when they had been bandmates in the wild with a stallion known as Apollo, who we had recently rescued with two of his sons (I will share their story next), and we knew it was important to keep them together.

↙ Stargazer, Soleil, Ladybug, Sedona, Hera, Strongheart, Firefly, and Isabella
↙ Strongheart and Hera

As both these incredible mares began to regain their physical strength, we decided reuniting them with Apollo and the sons he was fiercely bonded with wasn't what was best for them or him. Because of the two mares' "auntie" stature in the wild, we instead added Izzy and Wildheart to their group, and then two more: Ladybug had been rescued from a kill pen in Kansas, and her baby mule, Firefly, who was a real handful.

Firefly met his match when he met his "little sister" Wildheart, who, with so many aunties to dote on her, became quite the princess. She would regularly put her big brother in his place. Call it "mule karma," but Firefly has now settled into being a hen-pecked young male who puts up with his mother, aunties, and sister with so much grace and patience it's wonderful to watch.

So we have another family that might not all be blood related, but they are bonded and so close and we would never do anything to separate them now. Who knew when we rescued Ladybug and Izzy that we were actually rescuing four, not two, and that they would be the foundation for a whole new Skydog family.

—

J. Frank Dobie

The Mustangs

chapter

APOLLO, ASLAN, AND HERMES

18

18

The day that I traveled to the adoption event in Colorado where I finally saw and talked to Blue Zeus—and made my silent bid to adopt him—I walked the pens and looked at some of the other senior boys from the Red Desert Complex in Wyoming. They were in horrible shape—skinny, muddy, and beaten up. One horse stood out, because he didn't have the same cuts and bites on his hide as the others, although his face was badly smashed up, probably from being loaded and unloaded from the trailer, or in the chutes being processed and gelded. This was Apollo, the oldest of all the horses up for adoption at age twenty-two. I had seen photos of him free on the range, and he was far from looking like the same horse now. He was more afraid of people than most of the other horses and spent his time trying to hide with his son, Aslan, who was in the same small pen. They played a game of cat-and-mouse around and between the other horses, trying to get as far away as possible from nearby humans, while always keeping a watchful eye on them.

I didn't hesitate to adopt Apollo and Aslan, seeing how closely bonded they were. Then I discovered that another palomino pinto in a different corral (Hermes) was also Apollo's son. I put him on my list of horses to help solely to keep the family together. I thought if a father and his two sons could at least live

Apollo and Aslan when they first arrived at Skydog

out their lives together with familiar kin and friends, they might find some semblance of happiness. More than anything, I wanted to give these three the space they craved to get away from people, who, at the corrals, were obviously far too close for their comfort.

I only paid twenty-five dollars for each magnificent horse, as nobody bid against me. We arranged to bring the three to Oregon the same day, along with Blue Zeus. It was a dramatic and unsettling loading process, though, as Apollo was horribly upset. I watched helplessly as the Mustang tried to jump over fence panels and ran into gates, desperate to escape to freedom. He finally loaded, banged up and shaken, but we just were relieved to get him on the trailer heading home.

At the sanctuary, we unloaded Blue Zeus and Hermes first. The two geldings took off running into the forty-acre pen full of juniper and spring water, and as close to "wild" as we could provide for the first few weeks. We kept Apollo and Aslan back as we still had to take Apollo's tag off—we hadn't done so at the BLM corrals as management hadn't wanted to upset him any further prior to trailering. As nervous as we all were at the prospect of

pushing this huge wild boy back into the chutes he despised, he actually went in incredibly calmly and stood still as we cut off his neck tag and immediately let him out.

We quickly reunited Apollo and Aslan with Blue Zeus and Hermes. At first, Apollo and Aslan ruled the pen as quite the powerhouse pair, but soon they all settled into a routine just the four of them, very much two distinct pairs, and Hermes seemingly the lowest on the totem pole.

When I first adopted Hera and her companion Strongheart,

Aslan and Apollo

Looking at these faces we know we can't slow down or shut up. So many more people around the world now know the plight of the American wild horse, and we won't take this as a loss but rather as a step forward in a better future for them all. So take a moment up on this hill with these wild horses and breathe and cry if you want to. I can't tell you how many tears of frustration I have shed over the years for the ones left behind, fights lost, and watching families torn apart, and it's why we work so hard to reunite the ones we can. Because it matters. Family matters. Mustangs matter. Reuniting them matters. We will help pick up the pieces and do our part with pure hearts and good intentions. For the horses. And for all of you too 💔💔💔

← *Hermes*
↙ *Apollo and Aslan*
↙ *Hermes*

whose stories I shared earlier, I was excited to think that we might be able to reunite Apollo with his lead mare from the wild. But when we tried, Aslan "stole" the two mares and left Apollo alone, which just wasn't right. The mares had been happy together, and Apollo and Aslan had been happy together—so, we again separated them. As it turned out, this family story was one of friendship over love.

After we turned Apollo and Aslan out with the group Hermes had been running with on over a thousand acres, for weeks, Apollo and Aslan hid in the trees and ran away anytime they saw people. But time works wonders. After months of becoming accustomed to the comings and goings of staff at the sanctuary, the two began to hold their ground and face their fear of people. Finally, they were convinced that we mean them no harm and come in peace.

Today the father and his sons live in freedom together, restored to their beautiful selves in all their magnificent glory.

As you follow the tracks of the wild horse,
perhaps you'll agree that it deserves
a safe haven in the country it helped to build,
deserves the protections it once had and were
only recently unraveled; perhaps you may
have a greater understanding of the forces
that are contriving to wipe out our
loyal partner, the one in whose hoof sparks
this country was born.

—

Deanne Stillman

Mustang: The Saga of the Wild Horse
in the American West

chapter

LEGENDS OF THE FALL

19

19

IN these pages we have seen that sometimes just the bonds horses form being rescued and brought to Skydog together forms a family all its own. A "band of brothers" is what I call the next family of senior Mustangs. Like Blue Zeus, they were all rounded up by the BLM from the Red Desert Complex in Wyoming. And when I went to find Blue Zeus, I also wanted to choose some other "old boys" who would be hard to adopt out and give them freedom for all their remaining days at the sanctuary. Little did I know how much we at Skydog, and others, would fall in love with the Wyoming wild boys.

Standing near Blue Zeus in the small pen was a jet-black horse with such a long forelock it hung like a velvet curtain over his eyes and face, creating an air of mystery, but also deep despair. He was the skinniest horse in the corral and seemed so lost and defeated and sad, and we knew we had to help him get out of there. We named him Spartan for his brooding handsome self and to give him the name of the fighter he later turned into at Skydog.

When I shared a video of Blue Zeus in that pen, noting that he had been found and we were bringing him home, along with several other senior Mustangs, a group of eagle-eyed followers pointed out a brown horse with frostbitten ears that made him look rather like a teddy bear. His tag number, #5001, was clearly visible in the video, and the followers started a campaign to try to persuade us to adopt him, too. They dubbed themselves "The Sisterhood of Hope," in the hopes we would be able to bring one more home with us.

Well, in the end, we had a spot on one of the trailers, and we added the Mustang with teddy-bear ears as a surprise to all the ladies who had never given up asking for him to come to Skydog. Noble, as he was named, traveled home to us and took his place with the other "Legends of the Fall," as we came to call this special group of seniors.

A couple months later we returned to Cañon City, Colorado, with one goal in mind: to rescue four more of these incredible old boys and get them back to Skydog to live the rest of their lives free. It was one of the hardest things I ever had to do—go back

I can't imagine what it must be like for a wild horse who has spent over twenty years in the quiet of the wild, suddenly to have their senses overloaded by loud sounds and so many scary and noisy experiences. To see friends die. To lose your family and all you held dear. The most profound loss I can imagine suffering. So when Drifter took off alone I understood. I knew he was all right as I would see him on the security cameras and could spy him on the high hills grazing in silence. Just the wind and the bird song in his ears. And I understood. They all have to heal and grieve in their own ways and Drifter found his. With the scents of nature, the sage brush and pine in his nostrils, and the bright stars above, he healed. But today I spotted him with the boys and smiled. Drifter is back and I hope he's happy again. He didn't run when he saw us, and he exuded a calm acceptance of this new life 💙 Drifter you're a legend—welcome home 💙

Joker and King

into the prison and pick only four horses from the dozens of old Mustangs in front of me.

One of the corral managers helped me choose by looking at his list to identify the oldest and neediest four. A twenty-six-year-old grulla with crippled front legs that bowed and bent outward to the point of our not being able to figure out how he was still standing, let alone walking, caught my eye. I imagined him in his prime, out on the range and "king of the hill," with a family he loved and fought for. How this old horse had run from a helicopter and somehow made it alive to the corrals was miraculous. I truly did

not know how any horse in his state could have escaped the BLM euthanasia bullet, but he had. So I had to make it count. We named him Gandalf, as he seemed to be a wizened old spirit with magic in his eyes.

Then I chose another grulla who resembled him and stood close by, in case he was his son. A skinny dun named King was added, and a bald-faced black boy we named Joker. I filled out the paperwork and left, not knowing it would be the last time I would see the rest of the old Mustangs in the pens there.

On my way to the airport, the most spectacular full moon came up as I was musing over what to call the other grulla boy. I found out the moon was called the "Frost Moon," so that was the name that fit him perfectly.

I am thankful these senior boys, our Legends of the Fall, who lived their whole lives on the range with families of their own before losing everything they knew and understood, could be given back their freedom and wildness. That had to be enough for us. Even though I relentlessly called and emailed later, asking to be allowed to return and adopt more of the oldest, neediest boys, they were all sent to long-term holding or euthanized. There were no more there to help in Colorado. So many beautiful horses, lost forever.

*But among the many tribes that acquired
the mustang, it was the Plains Indians
who became the centaurs of the American
frontier. They called the horse sunka
wakan—sacred or mysterious dog.
'Dog' because it became the new pack animal,
replacing the smaller coyote-wolf crossbreed
that had served the Indian for thousands
of years, and 'sacred' because it was much
more than a carrier of goods: it was a hunter,
a warrior, wealth and prestige;
it was medicine, it was magic, and above all,
it was allied with the Thunder Beings
who lived in the west, where rain begins.*

—

Deanne Stillman

*Mustang: The Saga of the Wild Horse
in the American West*

chapter

THE MAINE THREE

20

And this is a story of three horses who were about as far as you can possibly get geographically from Skydog Ranch and Sanctuary and the journey they made to come home.

We received an email from a wonderful organization, Maine State Society for the Protection of Animals (MSSPA), about three Mustangs they had taken in from a terrible starvation and hoarding case in their area. The Mustangs were part of a group of twenty horses seized by animal control; all were in terrible shape and condition. Several had to be euthanized to end their suffering.

The three wild horses were originally from Saylor Creek HMA in Idaho, and up until that point, we had not yet taken in any Idaho Mustangs. The rescue had made efforts to gentle the three, but none of them were good candidates for domestic life, and so it was determined they would be best off in a sanctuary. We balked at the huge costs of transporting them across the entire country, from Maine to Oregon, but unbelievably, the MSSPA offered to pay for the haul, as well as send one of its staff with the horses to make sure they got to us safely. With two rescues able to work together for the good of these wild horses, it was impossible to say no.

We waited until the spring came, and a wonderful hauler with a huge trailer that had three big compartments—like stalls—so the horses could lie down and turn around during the journey was arranged. Because they were not gentled, they would have to stay in the trailer for the whole journey, so ensuring they had room to move was important to all of us.

A local NBC station, News Center Maine, and the wonderful reporter Peggy Keyser and photographer Kirk Cratty, had followed the story of all twenty rescued horses, and the station posted updates every day the three Mustangs were on the road, on their way back to the wild life they deserved.

Each day, as the horses got closer, enormous excitement built. I had never seen Idaho Mustangs, and it was exciting to know I would soon meet some. Various people learned the story and sent us photographs of the horses free on the range—Silver appeared to have been the leader of one of the biggest groups of horses we had ever seen, and the mares Sienna and Raggedy Ann were recognized by our followers, as well. It was incredible to gather so much information about their herd in advance of their arrival. The photos of what we called "The Maine Three" during the first days of their rescue had been heartbreaking, and so we

willed them across the miles and state lines so we could fatten them up and get them back to good health.

Finally, the day came when they arrived. It was beautiful to see them jump out, one after the other, into a large pen where they could decompress and settle. Silver had the big jaw and stature of a band stallion who commanded respect and attention. He was a sweet, loving, and attentive mate to the two beautiful mares he was with and obviously doted on them. Silver had some issues with his eyes, and we treated them every day with his cooperation,

Silver, Raggedy Anne, and Sienna

 **skydogsanctuary**

There are days that are hard and the overwhelm and burden of responsibility and level of hard work are difficult. The hopelessness that can creep in, the wanting to do more and feelings of not doing enough. The realization you can't save them all can take you to a dark place in rescue. When the problem seems insurmountable and the progress seems slow and the cruelty immense. It's moments like that when I have to step back and look at a family like this one and know it's worth it and that we are making a difference. I sit on a windy hill and look at how at home this family feels, comfortable and safe enough to lay down and nap. The slow rolling and sleepy casualness of feeling secure enough to take your time. And there's no sense of fear, no anxiety in their eyes, no worry or panic. Just peace. And we are honored that this is their home now forever, side by side.

as if he knew we were trying to help him. He wore a UV fly mask for several weeks as we acclimated the group, and then he and the girls were ready to be released on over a thousand acres with other Skydog Mustangs.

It was incredible to see The Maine Three run free. Silver immediately took the mares high up on a hillside and hid in the trees, as if he worried they would be caught again.

A few weeks later we released three new Mustangs—Hedy Lamarr, Saint, and Lark—and they joined The Maine Three to form a group of six with two former band stallions (now geldings) who pretty much keep to themselves other than at feeding time. They claimed territory near the Aspen grove

↑ *The Maine Three*

→ *Raggedy Ann and Sienna*

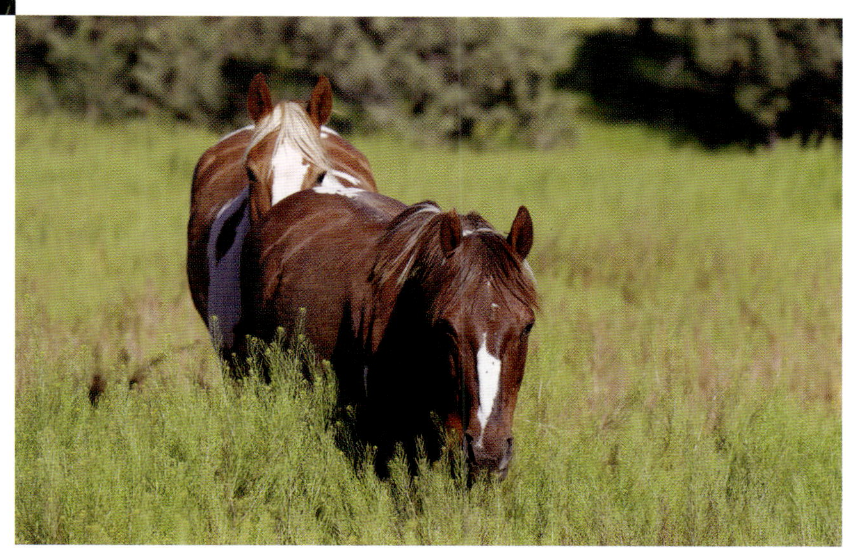

where a stream bubbles through, spectacular in their beauty and wildness. All six will happily take a cookie when thrown to them, but they prefer their distance and space from humans… just in case.

And so, another family joins the Skydog herd. Each time this happens, everyone who cares about wild horses—cares about and believes in the deep bonds they form; cares about love, compassion, and kindness; cares about the work we are doing and the families we reunite—breathes a sigh of relief. Another wrong has been righted as we continue to do all we can to bring peace and freedom back to those Mustangs taken from the range. After all these horses go through,

from the moment they lose their freedom and families to their time in holding facilities and kill pens, it is incredibly meaningful to see them living as they were born to be—wild and free…and together forever.

I see them, vanishing, vanishing, vanished,
The seeds of grass shriveled, two pens of barb-wired property,
The wind racers and wind drinkers bred into property also.

But winds still blow free and grass still greens,
And the core of that something which men live on believing
Is always freedom.

So sometimes yet, in the realities of silence and solitude,
For a few people, unhampered a while by things,
The Mustangs walk out with dawn, stand high, then

Sweep away, wild with sheer life, and free, free, free–
Free of all confines of time and flesh.

—

J. Frank Dobie

The Mustangs

chapter

NUGGET AND JOSEPHINE

21

21

AS of the time of writing, our latest addition is a brand-new baby. Nugget is a beautiful palomino mare we saw at the BLM corrals, struggling to walk with a locked stifle. She was dragging her back leg, and it was hard for her to get to water and food with such a condition. We adopted her and worked to unlock her stifle and get her sound and out of pain.

As we helped Nugget regain her health, we started noticing her belly was growing rounder and bigger. She was pregnant! And so baby Josephine, who we nicknamed JoJo, was born at Skydog. After the heartache of seeing many cremello Mustangs euthanized at the South Steens roundup because of concerns about potential blindness, it was a blessing to see her smoky cream color and blue eyes. This was one horse with a pale-colored coat and pale-colored eyes who would be allowed to live a long and healthy life. Because even if she did someday go blind, it would not be a death sentence. We would take care of her, along with her mother.

It is time for more of those working for the Bureau of Land Management to start truly caring for the wild horses in their charge and stop seeing them as a "nuisance" or a "problem." Mustangs can be a link to nature and freedom, and a "West" that should remain as wild as the horses and other creatures who live on its lands.

We would like to think that one day there won't be a need for sanctuaries like Skydog to provide a soft landing for the wild horses who have been gathered from our public lands, often for no good reason, but until then, we will keep doing our very best to care for them and inspire others to care about them.

And on we go…

There is a great saying I go to when things are out of my control and scary. "Acceptance is the answer to all my problems today. When I am disturbed, it is because I find some person, place, or situation—some fact of my life unacceptable to me, and I can find no serenity until I accept life as being exactly the way it is supposed to be at this moment. Nothing, absolutely nothing happens in God's world by mistake, and unless I accept life completely on life's terms I cannot be truly happy." I believe horses are the most accepting souls of any I have ever met. They live in the moment and accept where they are at all times with grace and often gratitude. These beautiful families were trapped and taken from their homes, lost friends and family and their lives as they knew them. But today they have adjusted and accept and welcome their new life. It's a joy to see them.

epilogue

Eight Things You Can Do
to Help the Wild Horses of America

1. Follow us on social media *@skydogsanctuary*, share our posts, and donate to our fundraisers to help us, help more Mustangs in need. The more people who find out about and fall in love with wild horses, the better. The more we are supported, the more horses we can save.

2. Contact your state representatives, remind them how important the wild horses are to Americans, and ask them to co-sponsor the SAFE Act, which would help stop the transport of wild horses across our borders to Mexico and Canada to be slaughtered.

3. Volunteer at a local horse rescue or offer your talents to a wild horse organization and work remotely to help in any way you can. At Skydog, we have some incredible people behind the scenes sending thank yous, welcoming new patrons, taking photographs of our wild horses to share with the world, and donating items to our auctions to raise funds. There are many different ways you can get involved and help the cause.

4. Contact your local media and ask them to run more stories about horse and donkey rescues or the wild horses and burros of America.

5. Adopt a Mustang or encourage your knowledgeable, horse-loving friends to do so. Mustangs are the most incredible horses—smart, hardy, sure-footed, athletic, and soulful. You can rescue one from the slaughter pipeline easily, and the journey you will take with that horse, learning how to gentle him and become his trusted friend and partner, is the greatest journey of self-discovery you can take.

6. Follow other reputable horse rescues and sanctuaries, and share and post about their mission and message to help raise awareness and educate others. We can all do a small part to change things for wild horses when enough people know about the issues and more get involved in the fight for their right to freedom.

7. Research the subject. There are many great organizations doing good work to educate people about our wild horses, including the American Wild Horse Campaign (*americanwildhorsecampaign.org*).

8. Talk about wild horses. Whenever you can get into conversations about Mustangs and the issues they face, see if you can inform more people and inspire them to get involved. Tell them about Skydog Ranch and Sanctuary. Display our merchandise and stickers to provoke conversations about Skydog when you're out and about. All proceeds from the purchase of our merchandise goes directly to maintaining the sanctuary and rescuing more wild horses (*skydogranch.org*).

acknowledgments

I would like to thank and acknowledge the following people
who helped me along my photographic and literary journey,
in support of the horses of Skydog Ranch & Sanctuary,
and without whom this book would not be
the beautiful creation it is.

Photographers Jamie Baldanza, Steve Rymers, Shannon Phifer, Janelle Hight, Steve Paige (American Wild Horse Campaign), Larry McFerrin, Laura Leigh (Wild Horse Education), Rayna Zemel, and John Wheland, who all freely took and donated beautiful pictures of the horses while out in a remote location in all weathers.

Deanne Stillman, David Philipps, and Chad Hanson for their beautiful poetic words about Mustangs, which are scattered through the book.

My crew at Skydog, who work tirelessly to feed and care for the close to 300 equines we have rescued.

My husband Chris, without whom we would never have been able to start Skydog and make it the wonderful place it is today, where horses can be horses.

Our followers and supporters, who lift me up and keep me going with their amazing likes, comments, and emails when they're needed most, especially our Patrons, who make it possible to do this work and to say yes to more horses and donkeys when needed.

Our amazing publisher Trafalgar Square Books—Publisher Caroline Robbins, Managing Director Martha Cook, my editor Rebecca Didier, and designer Katarzyna Misiukanis–Celińska. Thank you also to Seamus Lyte for connecting us at the beginning of this wonderful collaboration.

And thank you to my amazing first Mustang Buddy, who taught me so much about being of service, about Mustangs, about family, and about myself.

Skydog Ranch & Sanctuary
would love to hear from you!

Please visit us at

SKYDOGRANCH.ORG

for more information about the work we are doing
and the ways you can get involved to help wild horses and burros.

We invite you to follow the stories
of our wild horse families on Instagram and Facebook

@SKYDOGSANCTUARY